Creating the Miniature Australian Stock Saddle
English and Western versions

KeriOkie Entertainment

Edited by Susan Bensema Young

ISBN 9780976756484 © Carrie Olguin 2007, 2020

Acknowledgements

To Susan Bensema Young, my editor and friend, thank-you from bringing intelligence, humor and enthusiasm to such a thankless task.

Other Model Horse Hobby Titles by Carrie Olguin and KeriOkie Entertainment:

How to Make Miniature Saddle Trees Clay Edition

How to Make Leather Western Saddle Trees

Tooling the Miniature Roper Saddle

Creating the Miniature English Saddle with Tree

How to Dressage for the Model Horse Arena

Creating the Miniature Cutback Saddle with Tree

Modern Sidesaddles for the Model Horse Arena

Arabian Horse Costumes for the Model Horse Arena

Australian Stock Saddle, English and Western

Pony Express Saddle for the Model Horse Arena

Pack Saddle for the Model Horse Arena

For more information about the author and publisher visit www.KeriOkie.com.

Copies of this or other titles can be ordered from:

Carrie Olguin
KeriOkie Entertainment
341 N Laura Dr
Chandler, AZ 85225

Practice! Practice! Practice! Few of us get it perfectly right the first time. When you critique your work, stay away from phrases based on whether you love or hate the work in progress (save that for after the project is complete or you will never get it done). Instead, critique the technique and not the results. I have seen too many people give up because they tell themselves how terrible the work is instead of concentrating on how to make it better. Every media (leather, paint, clay, etc) has a life of its own. It is up to the artist to shape (not control) the form, giving breath to that life.

Warnings

Children need to have parental supervisions when working with the tools and chemicals presented in this guide. Glues, dyes and other chemicals have warning labels for a reason. Read those labels and take the proper precautions recommended by the manufacturers. Scissors, razors and other tools, though useful, are also potentially dangerous. Use with care and under appropriate supervision. The author and publisher are not responsible for any damage or harm that occurs from the use of any substance, material, tool, chemical or other item mentioned for use while creating the designs or patterns in this book. The reader is strongly advised to know the potential harm that could occur and to proceed with care and caution at your own risk.

Author's Notes

It has taken me much longer that I intended to finish this book. So, I apologize for the delay. I am quite proud of this set of patterns. It is, however, one of the most complicated patterns I have so far created. I won't say it is easy or hard. I have too much experience now to make such a judgement. (Everything is hard until you know how. Then it's easy.) I will say that previous model saddle making will help tremendously and may even be required for successful completion.

If you have never made a model horse saddle before, I suggest you start with my cutback or dressage saddle patterns, as these are much less complicated. I certainly do not want anyone to become frustrated when making this saddle and decide never to attempt tack making again, thinking they are not capable. The reason why I invest the time and trouble making these pattern books is to share my love of tack making and spread my knowledge to as many people as possible before my hands and eyes get to old to make tack any more.

Some of my readers have hesitantly shown me saddles they have made from my patterns, somehow feeling guilty for making modifications. For those who have yet to approach me, let me assure you that I expect you to make modifications to the patterns to compliment your tack making style. I even expect that some of you will make better saddles than those saddles found in this book. I have three requirements for my patterns: 1) they must be repeatable, 2) the techniques must be teachable in this format and 3) the materials have to be readily available. Those three requirements do limit the quality of the saddles in this book. Not all of the tack that I make meets those three requirements (hence, you will not find those patterns in any of my books). Quite frankly, some of my best work is made of hard to find materials with techniques not teachable in book format. Therefore, I certainly do expect you to experiment and modify the patterns and with my blessing.

I was saddened to find out that some people are giving away copies of my pattern pages. This is wrong for many reasons, but two stand out. First, giving away the patterns without the rest of the book is like handing some one the schematics to a car engine and expecting a successful build up without providing the parts list or assembly instructions. That really does a disservice to the friend. Secondly, I invest a great deal of time and effort creating these books. I only get paid for that effort when a book sells. Without that financial reward, there is little incentive – let alone funds – for future projects or for me to continue creating patterns for our hobby.

Some of you have been watching this pattern develop on my web site, www.KeriOkie.com. I have a section under Tack called Anatomy of Prototyping a New Saddle Pattern that shows the entire process, even showing some of my experiments and certainly all of my failures. I hope this will enlighten everyone about the effort I invest to make these patterns available. I can guarantee everyone that I am not getting rich selling these books (too small of an audience). I am funding future projects and enjoying my favorite hobby. If you have techniques, concepts or opinions to share please feel free to write to Carrie Olguin, 341 N Laura Dr, Chandler, AZ 85225. I am always open to constructive criticism and polite critique.

Table of Contents

WHAT ISN'T COVERED 1

IN PERFORMANCE CLASSES 1

SUPPLIER INFORMATION 1

IDENTIFYING TYPES OF AUSTRALIAN STOCK SADDLES 2

FITTING AN AUSTRALIAN STOCK SADDLE 3

FITTING THE BRIDLES 3

TOOLS 4

SUPPLIES 6

MATERIALS 9

THE FITTING SADDLE 9

CUTTING OUT THE PIECES 10
    SKIVING 11

FAUX STITCH LINES AND TOOLING 11
    TOOLING PATTERNS 13
    REMOVING THE PATTERN PAPER 15
    TREATING THE EDGES 15

BLOCKING FORM TOOL 16

PRE – ASSEMBLY AND PREP WORK 18
    MAKING KEEPERS 18
    MAKING D-RINGS FROM JUMP RINGS 19
    USING THE BLOCKING FORM TOOL 19
    ASSEMBLING THE TREE 20
    PIPING DETAIL 25
    ADDING REAR FACING STUDS 26
    ASSEMBLING THE PANEL 28
    ADDING THE REAR FACING 31
    ASSEMBLING THE POLEYS 33
    ADDING THE POLEY PANELS 35
    PLACING THE D-RINGS 35
    ASSEMBLING THE WESTERN STIRRUPS 37

ENGLISH ASSEMBLY 40
    STIRRUP BAR AND BILLETS 40
    SKIRTS AND UPPER FLAPS 41
    FRONT OF SEAT PIPING 43
    STIRRUP LEATHERS AND FLAPS 44
    SEAT TO UPPER FLAPS 45
    FRONT PIPING - LONG PIECE 46
    PANEL TO FLAPS 47
    STIRRUP LEATHERS – FIXED 49
    STIRRUP LEATHERS - ADJUSTABLE 51
    SURCINGLE 52

- Web Style Girth 53
- Breastplate 54
- Saddle Blanket – Competition Style 56

WESTERN ASSEMBLY 57
- Stirrup Bar and Billet Flap 57
- Skirts and Upper Flaps 58
- Stirrup Fenders – Fixed 60
- Stirrup Fenders – Adjustable 62
- Upper Flaps to Lower Flaps 64
- Horn 65
- Horn to Seat 67
- Piping – Short Piece 69
- Piping - Long Piece 70
- Rear Rigging Rings 71
- Front Cinch 72
- Rear Cinch 73
- Front to Rear Cinch Connector Strap 75
- Surcingle 76
- Panel to Flaps 76
- Rear Billets 78
- Cinch Billet Adapter 80
- Roper Style Breast Collar 81
- Saddle Blanket - Western 83

BARCO BRIDLE 84

HALTER BRIDLE 86

CRUPPER STYLE 1 94

CRUPPER STYLE 2 95

TRAINING THE SADDLE 97

APPLYING A FINISH COAT 98

OPTIONS AND SUGGESTIONS 98

PUTTING IT ALL TOGETHER 99

SADDLE PIECES AND PATTERNS 100
- Reduction Ratios 100
- Blocking Form Tool 102
- Pieces Common to Both Versions 104
- English Version Pieces 106
- Western Pattern Pieces 108
- Saddle Blankets 110
- Faux Stitching and Tooling – English Skirts and Flaps, Rear Facing 112
- Faux Stitching and Tooling – Western Skirt/Flaps 114
- Saddle Bag Pattern and Assembly Schematic 116

What Isn't Covered

Putting ink to paper is an expensive endeavor, so I try to keep to the subject. Some of the things mentioned in this book are covered in other publications devoted to the subject of leather working. I try to keep to the information that is not easily found in other publications.

Dyeing leather – It is a simple enough process. But for all of the little details, you will need to refer elsewhere.
Tooling – Though I do provide tooling patterns, I am assuming you have tooled before. As the subject has been covered in many other publications, including one of mine; "Tooling the Roper Saddle."
Skiving – Refers to the process of reducing the thickness of the leather by scraping with a razor knife. Since this is an advanced saddle pattern, I am assuming you know what is involved and have had some practice at it.
Accessories – Aside from the rear saddle bags, a last-minute addition, I don't go into detail on how to turn this saddle into a police horse costume or endurance race entry. The details are left for you to research.
How to make miniature hardware – I don't include step-by-step instructions on how to make buckles, English stirrups, hooks, or any other specialty hardware. Quite frankly, making hardware is a bunch of work and the results are rarely the same as purchasing the correct hardware from Rio Rondo.
Braiding – Most people have mastered a three-strand braid by age 15. Braiding for the lead rope on the halter bridle is an option. The lead rope can be made from twisted cotton twine, yarn or even from leather. If you don't know how to braid, there are other options.

In Performance Classes

Australian stock saddles have gained popularity because of the comfort provided to both horse and rider. In the real world, Australian Stock saddles are used for mounted police, endurance riding, natural trail and an Australian sport called Campdrafting. For the model horse show ring that translates to Other Costume, Other Performance and scene or vignette classes. The use of an Australian Stock saddle may be limited in the show ring, as it is neither a Western nor an English saddle and cannot be used as a substitute for those classes. But it's the type of saddle that, when properly used, makes the entry stand out.

Supplier Information

Though I go through all of the tools, supplies and materials used in further detail in separate chapters, I thought it might be helpful to list the suppliers and web sites in one location.

Tandy Leather Factory – www.TandyLeatherFactory.com. Source for leather, tools, books and chemicals. They have just about anything related to making miniature tack except for Skiver leather and model horse hardware.

Rio Rondo – www.RioRondo.com. My primary source for model horse hardware such as buckles, stirrups, hooks and D-rings. They also supply small cuts of leather including Skiver leather.

Michaels – www.Michaels.com. General-purpose craft store. The web site has a store locator and a decent selection of items available for purchase on-line. They carry Tacky glue, craft tools, stick glues, paper embossing tools, jump rings, eye hooks, etc.

Joann's – www.Joann.com. General-purpose craft store. The web site doesn't have much of interest for model tack makers; but the store locator is helpful. Once in the store, there's a great selection of jewelry making supplies, hardware, thread, glues, etc.

Harbor Freight – www.HarborFreight.com. A really cool tool shop, which sells wire cutters, pliers, files, clamps, etc. Prices are reasonable.

Nicaven - http://www.nicaven.com. A discount supplier of scalpel handles and blades. A box of 100 blades is less than $10.00. Be sure to purchase the correct handle for the blades.

Identifying Types of Australian Stock Saddles

The Australian stock saddle started as an English saddle, but over the years has evolved certain characteristics that make it unique. The first, and most distinctive, characteristic are the poleys, braces for the knees and thighs that help keep the rider in the saddle, especially for mountain work. All Australian stock saddles have front poleys. The rear poleys tend to be optional. The second unique characteristic is a bar in the center rear of the saddle tree that allows for the use of a strap called a crupper (or croupiere) that extends from the center back of the saddle and wraps around the horse's tail. The crupper keeps the saddle from slipping forward. The third unique characteristic is the use of a fabric called serge used for the underside panels. It is suggested that serge panels provide more padding, better ventilation and less irritation for the horse than leather panels. The fourth unique characteristic is the use of a surcingle (or overgirth) as the second billet strap. The surcingle wraps over the top of the saddle, slipping through slits just behind the stirrups. The hybrid shown below has only one of those unique characteristics - front poleys. I consider it more of a modified Western endurance saddle than an Australian stock saddle. Construction of that form is not included in this book.

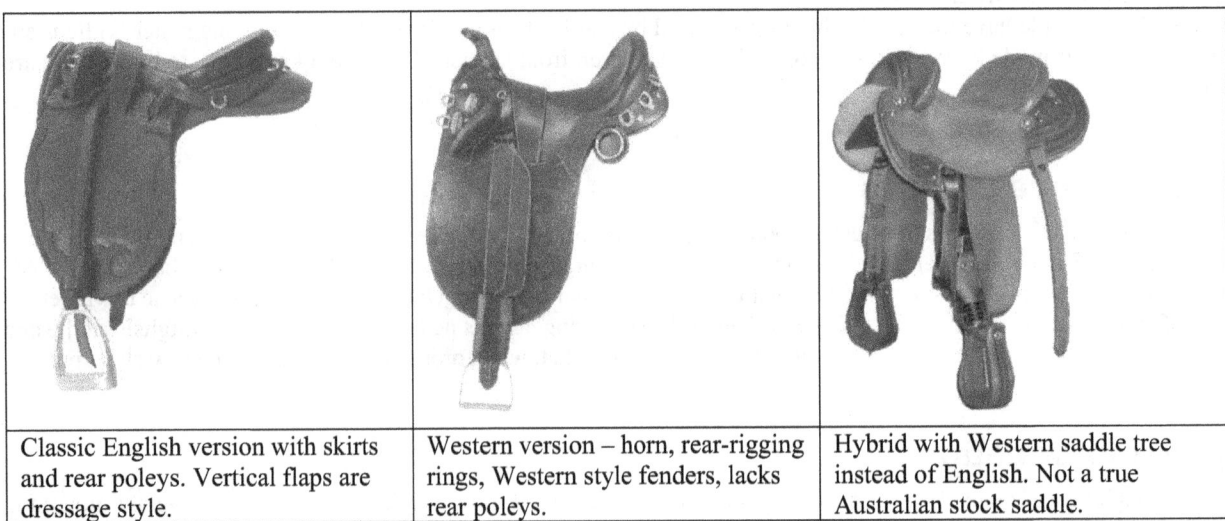

Classic English version with skirts and rear poleys. Vertical flaps are dressage style.	Western version – horn, rear-rigging rings, Western style fenders, lacks rear poleys.	Hybrid with Western saddle tree instead of English. Not a true Australian stock saddle.

Australian Stock saddles can be found in just about any combination of Western and English styles. The English saddle pattern in this book is based on the standard stock saddles developed in the 1930's that were popular with the bushmen due to ease of in-field repair. The Western version is based on the Somerset style developed some time in the 1970's and is the style most used in modern times. The basic difference is that with the English the skirts are separate from the upper flaps. With the Western, these two pieces are combined. The Western version can be made without horn, rear rigging and fenders, which essentially makes it a Modern English version. It can have either Western fenders or English stirrup leathers. Either type of stirrup is permissible. The same thing goes for the English version.

Though it would be more correct for me to call the English version the Standard Australian Stock Saddle and the Western version the Modern Somerset Australian Stock Saddle, the terms English and Western tend to be more familiar and descriptive for my purposes.

Fitting an Australian Stock Saddle

Though similar to an English saddle, the Australian stock saddle has vertical flaps. That means it fits more like a dressage saddle. Fitting any saddle is often a subjective matter. But the following rules apply to almost all saddle types. The front of the saddle should set over the base of the withers, far enough back so as to not interfere with the shoulder joint. The back of the saddle should never set on the loins. The girth should be far enough back so that it doesn't rub the elbow. The bottom of the stirrups should be flush with the bottom of the girth area. There should be an open channel along the spine from wither to loin.

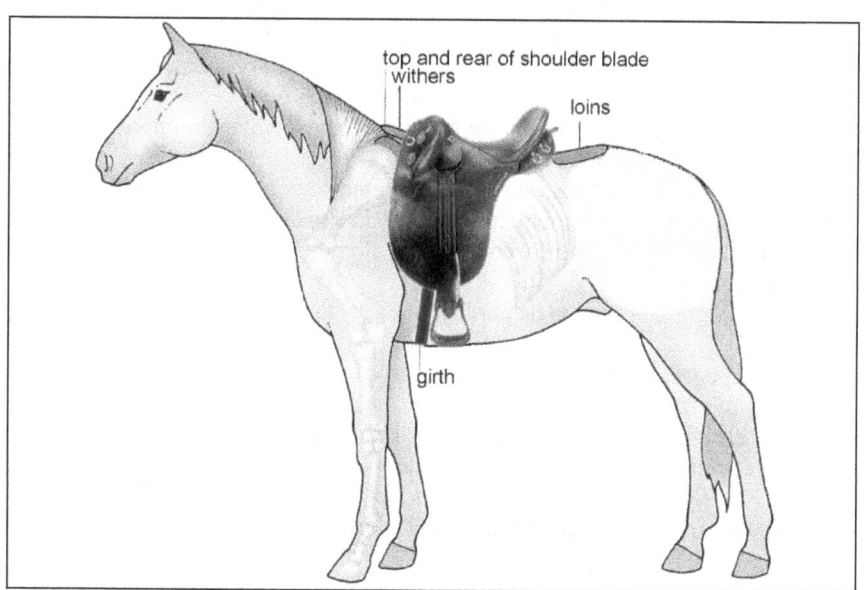

Fitting the Bridles

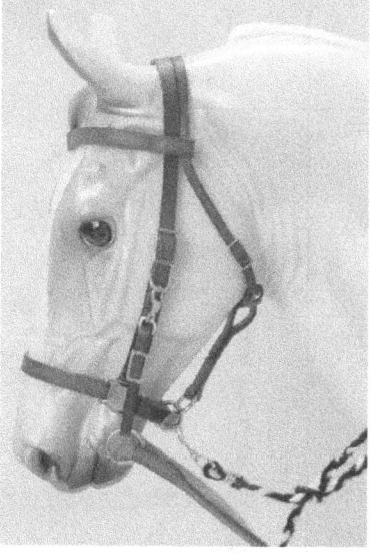

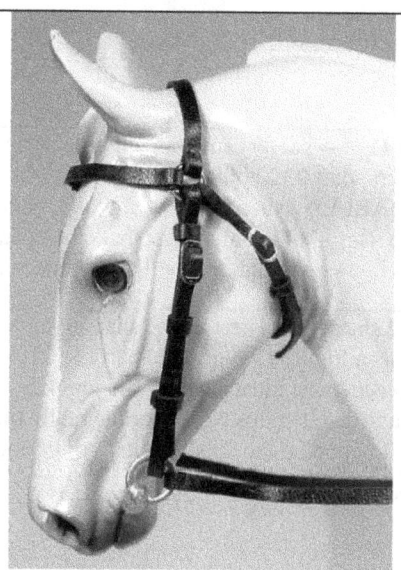

The halter bridle is used for endurance riding as it weighs less than the combination of both a halter and a bridle. The lead rope is necessary for the vet checks and to tie the horse at rest stops along the course.	The Barco bridle is another Australian invention and is the type used in the movie "The Man From Snowy River."

Any bridle type is acceptable for use with an Aussie saddle, but these are the two most commonly associated with it. As with all bridles, the junction of the browband should be low enough to allow free ear movement. The cheekstraps should not rub the eye socket. The throatlatch should be snug but not tight. For the halter bridle, the halter rings should be set above the bit rings and the noseband should be loose. Reins are normally the English style, though Western bits and reins are also acceptable.

Tools

Tools are the things you purchase (hopefully once) and use to make the project but don't end up as part of the project. You may have most if not all of these in your toolbox. If you have favorites, you will most likely prefer to stick with what you have. The point of this list is to show useful items that may help you achieve the best results possible. It's not a shopping list.

Scissors – This is one tool you should not skimp on. Get the best you can afford. Nail or embroidery scissors come in handy for small trim work. Leather cutting scissors really make a difference for cutting out leather. Tandy Leather sells a nice pair.

Sewing Machine – Though you can hand sew the panels, it is much easier to use a sewing machine.

Razor knife – Tester, X-Acto or generic. I prefer to use a scalpel, #11 blade. Scalpel blades are sharp. I have a special place for it on my shelf so that my toddler can't reach and so I don't cut myself searching my toolbox

Pliers – I suggest the craft size pliers, one pair with teeth (the teeth will leave permanent impressions on tooling leather) and one pair without. If you can only have one pair, toothed is better.

Wire cutters – For larger size wire (18 gauge and larger) and for snipping straight pins, you will need a good pair of wire cutters. Smaller craft cutters do a great job cleaning off joining points on the Rio Rondo hardware.

Rubber bands – Used for training the assembled saddle and holding aluminum to the sculpting surface. Thick bands work best. You could also use a shoe lace or ribbon.

Glue spreader – Toothpicks and straight pins tend to be popular choices. Glue control is one of the many important skills a mini-tack maker learns; that and skiving.

Paint brushes – For conditioner, topcoat and edge coat use a flat sable or camel hair brush, about ¼" wide. Use a disposable paint bush 2-3" wide, for applying dye to leather.

Body box models – Have one for each model size you intend to make saddles for.

Eye protection – Protect yourself from flying pin ends and other debris. A set of clear plastic safety goggles is a good investment.

Rags – To wipe the glue off of your hands. Old kitchen towels or T-shirts work well.

Round tip pliers - Found in the jewelry making section of the local craft store, these pliers are perfect for making your own cinch ring tongues as well as for shaping straight pins for Western style horns.	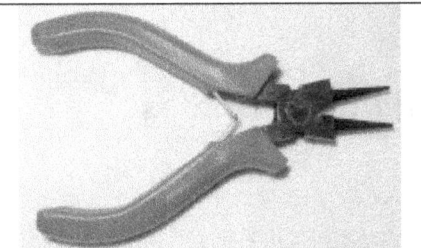
Punching Tool – Tandy came out with the perfect hole-making tool for miniature tack makers, part #3229-00. The largest hole is a 00! It has four additional holes SMALLER than a 00. Plus, it works on leather up to a 5oz. Okay, it's not cheap. But it's the only way I make holes in leather now that I own one. A 00-hole punch, awl or hand drill will also work to make holes.	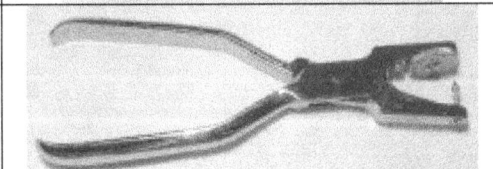
Stylus and shapers – There is one shaping tool I recommend you purchase. Tandy sells a Figure Carving tool part #8803105. Use it for embossing tooling leather. I consider it a must have for the tool kit. A small blunt point stylus works for the really tiny detail embossing work. It also works to open up the slits for the D-rings and surcingle.	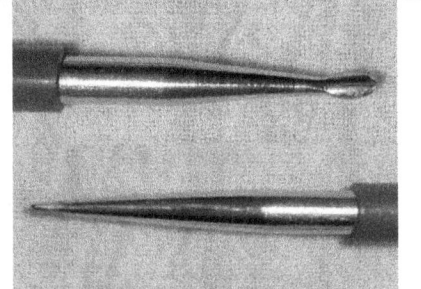
Clamps and clips - Look for small clips (small plastic or small clothespins) that don't have teeth and have a good spring mechanism. A dozen is a good number to have. A mix of types helps as well. The Micro Clamps pictured can be purchased from Harbor Freight (www.harborfreight.com), a pack of six for about $3. The small clothes pins (wood or plastic) can be found in the local craft store (check the baby shower supplies for some in pink and blue plastic).	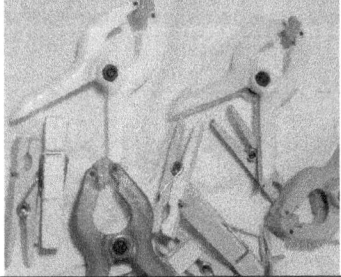

Dental tools – These can be found at hobby stores, DIY stores, flea markets, and Hobby Shops on the Internet. I like this spatula style for embossing the leather.	
Leather edge tool – Tandy sells a couple of hard plastic tools that you use to rub against the edge of the leather to give a professional finish. I prefer the tool pictured.	
Cutting board – I use a small wooden one. The thickness is great for getting the right angle when skiving.	
Marble surface – For carving, stamping and embossing, you need a hard, flat surface like a piece of marble. I have an 8" round trivet that I purchased from a kitchen outlet store (cost about $6.00). It handles water well and is dishwasher safe. If you don't have marble, you can use a piece of kitchen counter top material. Many DIY stores have marble tile for sale at a pretty reasonable price.	
Awl – This is a metal stick with a handle. There is only one requirement: Sharp! If it doesn't come with a point guard, make one out of cork or tubing and keep it on the tip unless in use.	
Magnifying glasses – For smaller saddles and detail work, get a pair of 3X (sometimes marked 300) glasses. Magnifying-hoops get in the way and are bulky to store. Since I only use these for craftwork, the case it came with protects the lenses from the other tools in my tool kit.	
Tin Snips - For cutting off the tops and bottoms of aluminum cans. Or you could try using a handsaw. Once the top and bottom is cut off, aluminum can be cut with a good pair of scissors.	
Files – For finishing buckles and stirrups from Rio Rondo, a set of jeweler files always comes in handy. Harbor Freight sometimes has a set of diamond edge jeweler's files that are quite nice.	

Supplies

Supplies are the items you use up in the process of making the project, end up as part of the project and need to be replenished from time to time. If you have made tack before, you probably have some of these already in your craft making kit.

Plastic bags – A simple way to keep your saddle pieces together.
Razor blades – Depending upon the razor knife you use you will need replacement blades. X-Acto knives are too dull. Testor's are better. I prefer #11 scalpel blades. Medical surplus supply stores are a good source of inexpensive replacement blades.
Paper plates – Use these to hold the waste materials, glue, and project materials and to keep your working area clean.
Straight pins – You want the pins *without* the ceramic or plastic balls at the end. You will use as many as 11 per saddle. The length should be 1 ½" for this saddle. The thicker pins work best for stirrup and horn assembly and the thinner pins for rear facing studs and final saddle assembly.
Jump rings – The most common jump rings used here are 9mm, 7mm, 6mm and 5mm, for bits, breast collars, D-rings and cinch rings.
Wire – For making cinch ring tongues, use a 20 to 24-gauge silver wire (The bigger the number, the smaller the wire). I also use straight pins or eye pins.
Paint – I use standard acrylic paint found at the local hobby store. Tandy has premium acrylic paint for sale but I've seen no difference except for the price. The two colors used in this book are black and white.

Glue – Fiebing's Leathercraft Cement is a flexible, non-flammable, no fume white glue specially formulated for leather. It sets up fast on a leather-to-leather bond. Tandy Leather (www.tandyleatherfactory.com) is a good source. Tandy has their own brand, which is just as good. Tacky glue is a good substitute. It isn't as wet as the leather glue and actually works better with craft felt.	
Gel Glue – To bond metal to leather, you will need stronger glue. Use gel instead of liquid for better control. Epoxy will work, but since you don't need that much glue, mixing epoxy correctly can be a problem. A warning: after years of exposure to chemicals such as this glue, I now develop a rash if it gets on my skin that doesn't go away unless I go to a doctor for medicine. I always use gloves when I use this glue and I don't use it often.	
Glue stick – Kid Stick and Elmer's School Glue Stick. There is also a Scotch brand that works well. Rose Art tends to dry up in the tube too quickly. The Sticky Note type glue doesn't work at all. Even if it says it's a permanent glue, that isn't true for a leather to paper bond. Target, craft stores or an office supply store are good sources.	
Edge coat – Professional saddles always have finished edges. Fiebing's makes an edge coat (black or brown) with or without an applicator. You can use a small craft brush for better control. Tandy Leather has Fiebing's and their own brand. Sometimes I use a Sharpie instead of an edge coat.	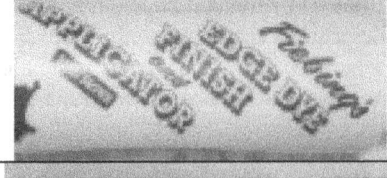
Pens – I only use a pen when I absolutely need to. The permanent ink in the Sharpie means it won't rub off on the models or my hands once it dries. I use the fine point to help prevent bleed through.	

Antique Finish – Fiebing's Antique Finish is a thick dye paste applied AFTER tooling to highlight the low spots. It is simple to use and the effect is wonderful. Apply a resist coating (Satin Sheen) before applying the paste to minimize the staining effect.	
Leather conditioner – Leather has natural oils that make it flexible and supple. Water and dye remove much of this oil. A good conditioner puts the oil back. I use Neatsfoot oil, another Tandy Leather product. Apply all over and on both sides (including the bridle and breast strap) 24 hours before applying a finish coat. It can be applied before shows to keep the leather supple. It does darken the color of the leather. But it adds a depth and richness that real "well oiled" saddles have.	
Finish Coats – Satin Finish provides a somewhat matte finish and is also used as a resist medium (resists dyes and stains, not paints) for dual colored finishes. Super Shene provides a shiny finish unsuitable for this type of tack. Tandy or craft stores carry these.	

Buckles – You can make your own out of wire by bending it around the tips of pliers. If you do make your own, I recommend soldering the ends.
    I prefer to purchase my buckles from Rio Rondo. Buy the buckles that match the width of the leather lace you will be using.
    Cast buckles are more expensive and more realistic. All Rio Rondo buckles require some cleaning. The cast ones often require polishing.	

Buckle Tongues – You can make these out of straight pins, 19 to 22-gauge wire, or purchase the pre-shaped type from Rio Rondo or your local craft store.	
Hooks – Rio Rondo makes these hook styles. You could twist some wire and make your own or use jewelry clasps. But these are perfect in scale and design. Note: both hooks still need to be cleaned of the excess connector metal on the top/bottom.	

Halter Rings – These are another Rio Rondo product. The Traditional scale rings are good for 1/8" lace and the Classic scale for 1/16" lace. You can substitute 6mm jump rings.

Note: The left ring still needs to be trimmed. The right ring has been trimmed.

D-rings – For this saddle, and any English saddle, if you can afford to buy the Rio Rondo D-rings, do so. For the pommel and rear facing use 1/16" or 1/8" D-rings, depending upon the size lace you will be using for the breast strap lace. You can make D-rings from jump rings as an inexpensive substitute.

Note: those pictured still need to be clipped from the excess metal.

Sticky Wax – for keeping bits in place. Found in craft stores or from Rio Rondo.

Materials

Materials are what the project is made of and will determine the quality of the final results. Leather is the main ingredient to saddle making. Thickness is noted by an ounce rating system; the smaller the numbers the thinner the leather. A 2/3 oz or "two-three ounce" is much thinner than a 7/8 oz or "seven-eight ounce". There are two main types of leather: tooling and full-grain. For realistic miniature saddles, you should use tooling leather. For practice and fitting saddles, you can use full grain leather.

Tooling Leather – Tooling leather comes in natural (not dyed) vegetable tanned. Tandy has tooling leather that is 1 ½/2 oz that works great for English saddles, called petite calf. It's a bit pricey and tends to have many flaws, but the skins are small, making it possible to purchase an entire skin. You will, of course, have to dye this leather to the desired color. For this hide, use the professional dye, as these tend to give better color. Tandy and Rio Rondo have a 2/3 oz tooling leather that is also suitable for this type of saddle. It is less expensive and stiffer. The hides are larger with fewer flaws. Rio Rondo has bags of scrap that are suitable for practice saddles and even for Live show quality saddles if the pattern pieces are placed properly.

Skiver - For the seat, piping detail and panels, use what is called skiver leather (lambskin). I call this leather a 1oz throughout this book, but that is not entirely accurate. Rio Rondo offers small pieces of dyed skiver in black and brown, just in case you don't want to dye it yourself. There really isn't a substitute for this leather. Tandy discontinued carrying this product. But I have found some for sale on eBay, though the quality can vary widely and the thickness is not always uniform. Look for vegetable tanned 1 – 1 ½ oz Skiver or Pliver. Chrome tanned (dyed) will work if you can't find the vegetable tanned. But you will be stuck with the color.

Leather Lace – Available in colors and natural. The sizes you will use is 1/4" for stirrup leathers, 1/8" for bridle and breast straps and a little bit of 1/16" for the halter bridle and the crupper style 1 pattern. With practice, you can cut the 1/8" in half to make 1/16" lace. If you purchase natural lace, you can dye it. Animal type (kangaroo or calf) doesn't matter as long as the grain pattern is small. Kangaroo tends to be better but calf is easier to find. A flat edge is easier to skive than a beveled edge. Rio Rondo has the best lace supply.

Aluminum can – This is an excellent, inexpensive, readily available material. You may need a pair of tin snips to cut off the top and bottom of the can. A good pair of craft scissors works to cut the saddletree support piece.

Fabric for saddle blankets – The type depends upon the blanket you want to make. Velvet or fabric with a high nap can substitute for wool fleece, as can chamois or deerskin suede. Tightly woven upholstery fabric works for event jumping and trail blankets. Check out some of the tack stores or web sites for the different fabric choices before hand to make certain what you buy is suitable. Craft felt is just right for felt type working saddle blankets and is what I used for the Western pad.

Craft felt – Yes, that inexpensive stuff you can buy in 8 ½" x 11" squares for less than $1 each. There doesn't seem to be any rule about what colors are allowed for the serge fabric. Red, green and blue seem to be the most common.

Elastic – For a web style English girth, elastic is a good material choice. I use ½" for Traditional size models. White is just about the only available color. But you can dye the all cotton variety if you want a wider color choice.

From this point on in the book, it is assumed that you are using leather that is already dyed and the hardware is already prepared. The saddle shown in the assembly pages is made of skiver, 2-oz tooling leather, craft felt and aluminum can. For brevity, substitutions of these items will not be discussed.

The Fitting Saddle

Every time you acquire a new pattern, you should make at least one "fitting saddle" to see how it fits the model and how it fits your tack making skills. Thin full-grain leather works well for fitting saddles. It's also a way to get a saddle made quickly, and get you excited about making another. You can use seconds or flawed tooling leather. Just don't use your best stuff for your first saddle. You will make mistakes. That is a normal part of the learning process. Don't expect the fitting saddle to be Live show quality work. That is not the point. The point is to get to know how the pattern fits your tack making style. When complete, the fitting saddle should show you the suitability of the tools, supplies and materials used, and allow you to make adjustments.

Cutting out the Pieces

Tools: Scissors, razor knife, and cutting surface.
Supplies: Stick glue.
Materials: Pattern, pattern leathers, and craft foam, aluminum and saddle blanket fabric.

Leather is a product of nature and therefore has flaws. Small flaws look huge on a small saddle, so your first objective is to locate the flaws in the leather and do your best to piece the project to fit around those flaws.

The oldest way to cut out a pattern is the trace method. If you are used to that technique, then by all means, do it your way. However, if you decide to trace, be certain to CUT AWAY the trace lines because tracing can increase the finished size of the pattern and you really do not want any pen or pencil residue messing up your finished piece. If you use a pen, a thin line is better than a thick line. For darker leather, you might want to invest in a dark fabric quilting pencil (at your local fabric store) or a silver colored Sharpie.

I recommend the glue stick method, that is, *glue the pattern piece to the suede or wrong side of the leather and cut on the lines*. This requires a new copy of the pattern each time. A photocopy is fine.

1. Start with the blocking tool pieces if you intend to use this method (I highly recommend that you use this method). <u>**You only need to make this tool once for each saddle size**</u>.
2. Cut out the saddle blanket fabric.
3. Move onto the 1-oz pieces. Use the stick glue to apply the paper pattern to the suede side (wrong/back). For the Poley panels, you might find it easier to glue the pattern piece to the grain side for the faux stitching details. Be certain to cut out the holes for the Poleys.
4. Next, cut the 2 oz-leather pieces. Use the glue stick method if at all possible. The paper only needs to stick long enough to cut out the pieces, to skive and make holes. Keep the cutting edge of the blade (scissors or razor knife) at a 90-degree angle to the leather. I cut out what I can with scissors and then use a razor knife to clean up the details, make the slits, score lines and create holes.

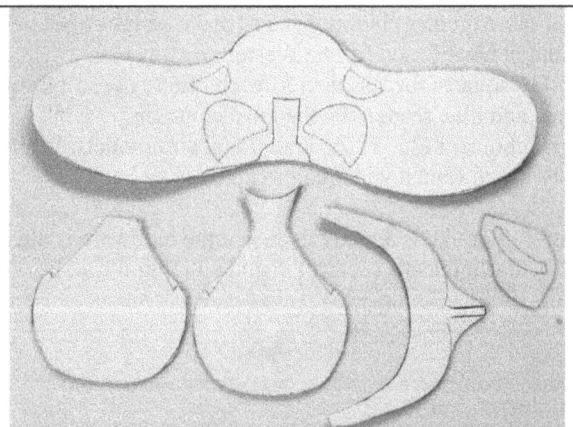
In-common pattern pieces that require razor knife work.

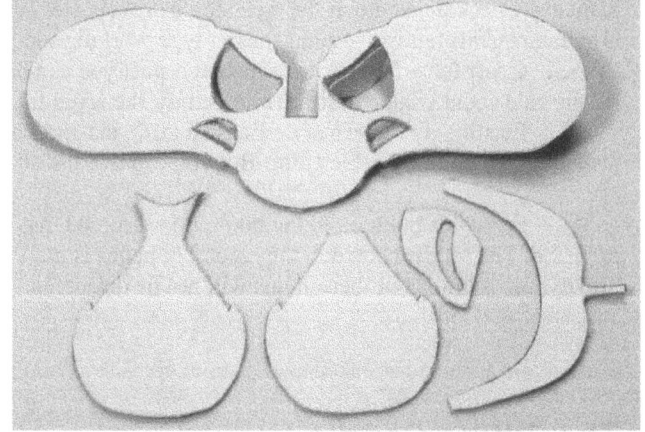
In-common pieces completed.

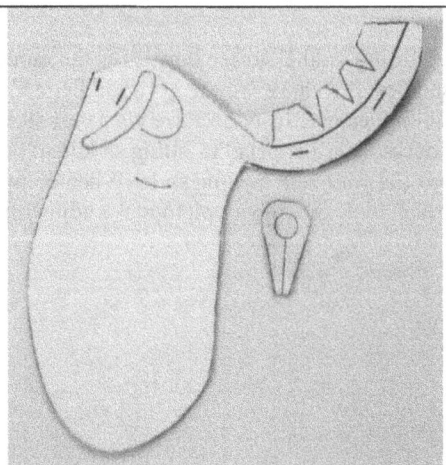
Western Pattern pieces that require razor knife work.

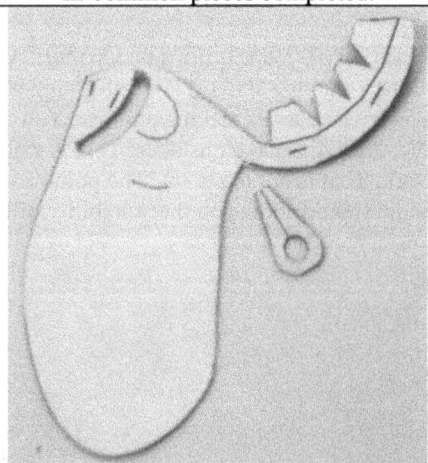
Western pattern pieces completed.

5. Finally, cut out the tree pattern from the aluminum. Take advantage of the natural curve of the can when placing the piece. You want the front of the seat to curve upwards.

Skiving

Skiving is the process of using a razor or blade to remove some of the wrong side of the tooling leather (or leather lace) to reduce bulk. Learning how to skive (and what to skive) will impact your entire tack making, not just this saddle. It is a skill that takes practice and not something that cannot be easily taught in this format. Start by removing small amounts in small strokes. Practice on scrap leather and lace until you get a feel for how to hold the blade. Some leathers are easier to skive than others. A sharp blade is essential. It helps to use the edge of the cutting surface to help get the correct angle without the handle getting in the way. Always skive with the blade stroking AWAY from you (and your fingers) to avoid injury. Skive the gray shaded areas on the pattern. Curved pieces are more difficult to skive than straight pieces. But the end result is worth practicing. Skive before soaking in water. Wet leather stretches easily and can be slippery.

Faux Stitch Lines and Tooling

Tools: Awl or needle on a stick, razor knife or dental spatula tool.
Supplies: None.
Materials: Upper flaps, skirts, cinch or girth ends, western stirrups, breast collar and rear facing.

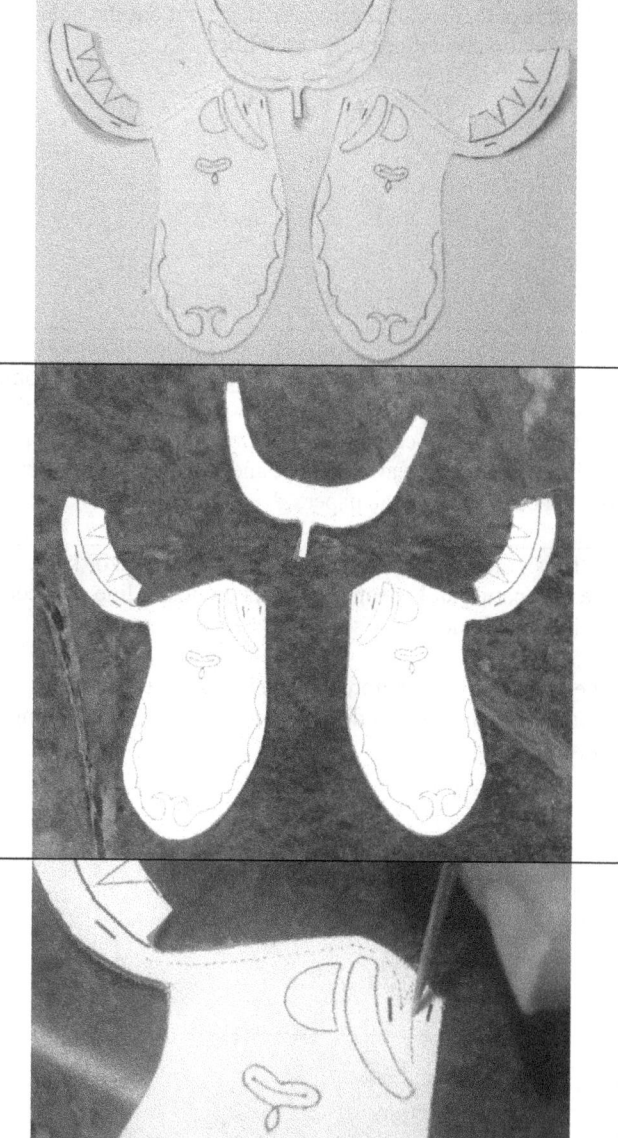

One of the simple details that makes a big difference in the finished work is faux stitch lines. Full size saddles are stitched as well as glued together. Since sewing on this scale doesn't work (to get the stitches small enough to look real, the leather would actually break. I've tried it.) I use another method to simulate the stitching. On a real saddle the stitches are about 1/8" apart. Any further apart and the saddle would not hold up to stress. Pictured here are the tooling guide pieces printed on standard 20lb paper and cut out.	
Use a tooling surface for this part. Using stick glue, glue the tooling guides to the RIGHT side of the matching pattern pieces. Use a light layer of glue, just enough to keep the paper to the leather for a short amount of time. **<u>NEVER pull the paper off of the right side of finished leather. It will ruin the finish.</u>**	
Use a needle, pin or awl and poke small holes along the stitching guide lines. Keep the holes close together, less than 1mm and keep the distance between the holes as consistent as possible. This does take a bit of practice, but it isn't that difficult.	

Now is the time to score a line around the back flap of the skirt. Even though the Western piece is shown, the same scored line is needed for the English pieces as well. This will allow the pieces to bend properly when glued into place. Score no more than half way through the leather thickness.

Do the same faux stitching technique on the rear facing piece.

The front poley supports on a real saddle would be sewn into place. Use the guides provided to add faux stitching, or soak and then emboss around the top, sides and part way around the back, and then all the way around the poley hole. There wouldn't be any stitching on the back part, as that is where the stirrup leather emerges from the stirrup bar.

If you decide not to use the rear poley, but do want to use the panel (which is actually quite common), don't cut out the hole for the poley but do still emboss the stitch lines.

For the rest of the faux stitching, as well as to complete the tooling, soak the following pieces in clean water. (If you plan to use my tooling patterns and techniques, finish scoring the pattern before soaking). This is also when I wet the seat cover piece for use with the blocking tool. It is important to soak the pieces that have the tooling and faux stitching guides. Let the pieces soak until the paper floats away. Rinse and rub the pieces to remove any glue residue. You might find it helpful to soak the aluminum tree piece to remove the paper.

To finish the Faux stitching, use the back of a razor knife (I like to use the spatula style dental tool) and emboss a line connecting all of the pinhole dots.

The top piece is the before and the bottom piece is the after. It is a subtle difference. But it does help to complete the stitching illusion.

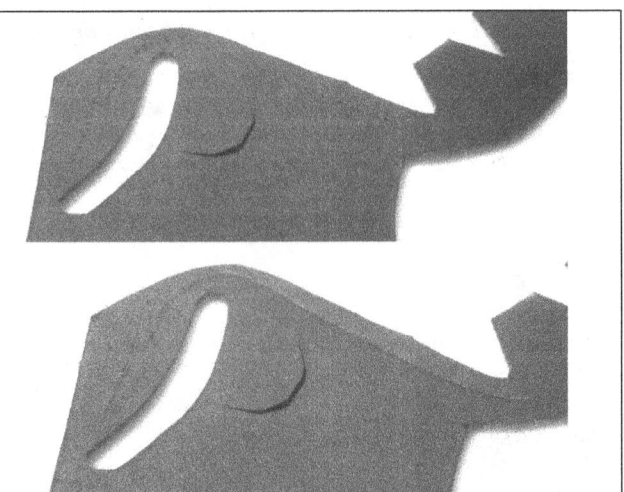

Do the same for the rear facing piece.	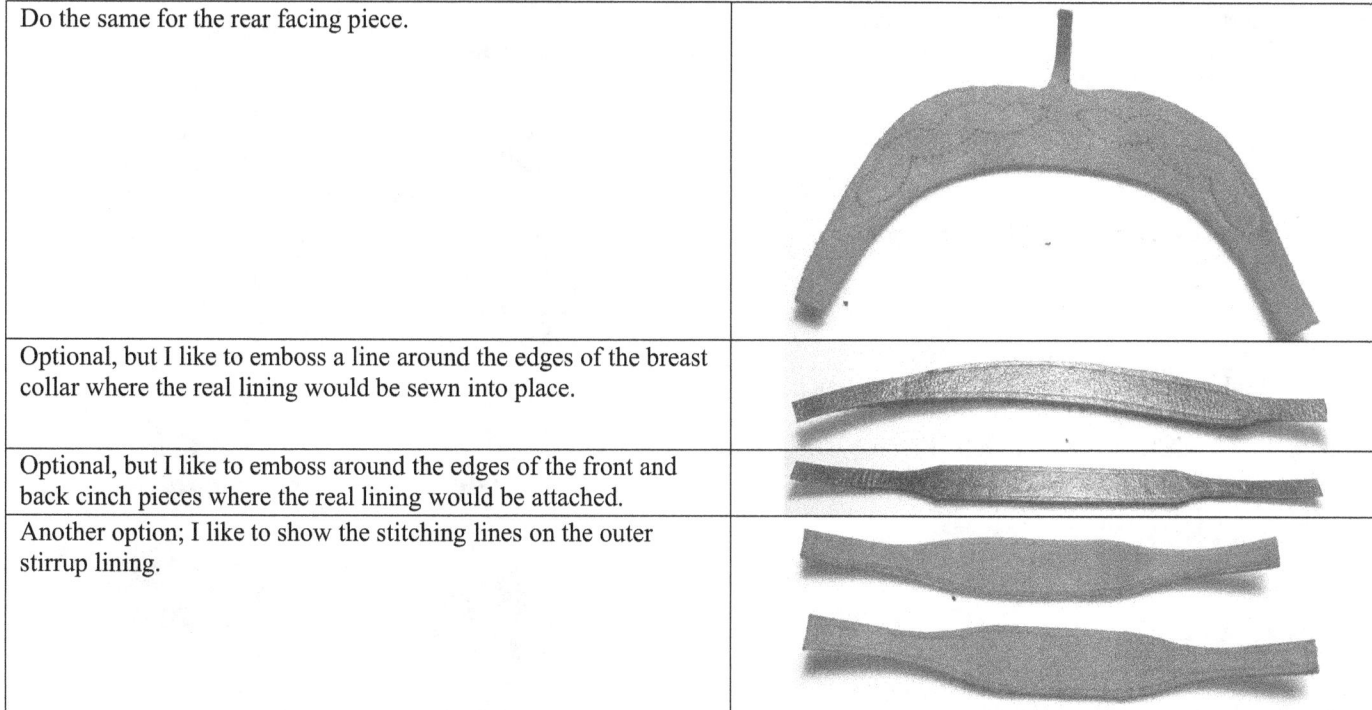
Optional, but I like to emboss a line around the edges of the breast collar where the real lining would be sewn into place.	
Optional, but I like to emboss around the edges of the front and back cinch pieces where the real lining would be attached.	
Another option; I like to show the stitching lines on the outer stirrup lining.	

Tooling Patterns

Tooling an Australian stock saddle is entirely optional. I am assuming you have some tooling knowledge. If you have never tooled before, I recommend looking into a class or a good Al Stohlman book. I cover the basics of tooling in my "Tooling the Roper Saddle" book. I don't plan to repeat all of that information here.

I include four tooling patterns. If you have used any of my Western saddle book patterns, you should be familiar with my technique. I use stick glue to adhere the pattern to the right side of the leather, just like for the faux stitching. Carve (score) the solid lines. Soak away the pattern paper (NEVER peel paper off of the right side of the leather or you will ruin the finish!) then let the leather dry for a while. I prefer to wait for drier leather before embossing and stamping. As with a soaking wet sponge, when you press down on the soaking wet leather, the water is squished away, but returns as soon as you release the pressure. With less water in the leather, less pressure is required to tool and the water does not return to the pressed area. As the pieces dry, the carved lines become easier to see. I prefer to work the leather when the top is basically dry and the middle of the leather still damp.

I believe results are more important than the steps used to get there. So, I don't disagree with other tooling techniques. I just find my way faster and easier than using a waterproof tracing paper method. I use a razor knife and not a swivel knife. A swivel knife not only cuts, but also opens the cut in the same pass. I have yet to find a swivel knife blade small enough for the tooling patterns that I create.

I use the word carving as another word for scoring the leather with a razor knife. Score along the solid lines, no more than half way through the leather. For this pattern, the heart piece around the surcingle slit is optional. Once the surcingle is in place, the heart doesn't look like a heart any more. But it is part of a traditional Australian tooling pattern. This is also the best time to make the slit inside the heart, to make certain it is correctly centered. Finish both flaps then soak away the paper and rinse off any glue.	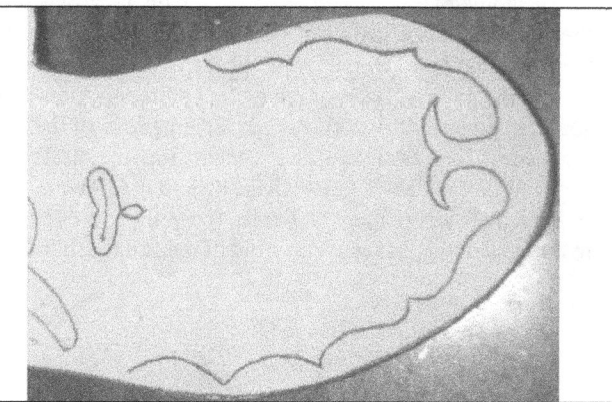

Let the leather dry for a while. The time will vary depending upon heat and humidity. The optimal time to tool is when the surface is dry but the center is still moist.

With this simple tooling pattern, emboss the inner edge all the way around the outside pattern edge.

Emboss around the outer edge of the heart shape.

After the leather has dried, an optional detail is to paint the raised areas. With this tan and black saddle, I used black paint. Any complimentary saddle color would work. I've also used a dark brown paint on a medium brown saddle.

This is another simple tooling pattern. The circular impressions around the outer edge were made with a paper embossing tool. I embossed around the inner edge of the two carved lines. For the three stems, I embossed on both sides of the carved lines.

This pattern is complex, but not as difficult as it looks. Carve the lines, then emboss around the outer edges. For the circle, emboss the inner edge of each line towards the center of the circle.

After tooling, paint the leaves and the medallion. Here, I used white. Any complimentary or contrasting saddle color will work.

For additional depth, use an antique finish. For this brown saddle, I used a mahogany color. Since the antique finishes change the color of the base leather, it may be necessary to use the antique finish all of the saddle pieces, tooled or not, to obtain a consistent color on the entire saddle. Or apply a resist medium (Satin Sheen) before painting help resist the staining effect.

This is the most difficult of the tooling patterns. After carving and soaking, press in the outside edge of the pattern all the way around the piece. Then form the flower centers. Playing the What's in Front, What's in Back game, make certain the flowers are in front of the leaves where there is overlap (push the leaves back) and make certain the leaves are in front of the rail (push the rail back).

Removing the Pattern Paper

If the paper resists removal from the wrong side of the leather you can either scrape the paper away with a razor knife or soak the pieces in water until the paper falls off. The glue stick will come off with water. It's not permanent glue (no matter what it says on the tube). If you need to soak, let the leather dry completely before moving on to the next step. Depending upon where you live and what the weather is like, that could be around 24 hours. How easily the paper comes off depends upon what brand of glue you used and how much of that glue you used.

For full grain leather and tooling leather, the paper tends to fall off pretty easily so soaking is not always necessary. Never soak suede. Always test a sample of full grain leather for water tolerance before soaking cut pieces. Some full grain leathers don't fare well, shrinking up and becoming stiff. This is especially true if using a Chrome tanned skiver, which is technically a thin full grain leather.

Treating the Edges

Take a look at a finished piece of leather, whether a belt or purse or billfold. Notice that the edges are clean, smooth and colored. If you look at your saddle pieces right now, you will notice that the edges are rough, possibly stringy and multi-colored.

To get that finished edge, you first need to rub the edge against a piece of hard plastic. Tandy makes a couple of tools for edge finishing. These used to be made of bone, but not any more. I prefer the stick style as the opposite end has ridges that work well for the 2 oz leather. If you don't want to invest in a tool right away, you can use the side of a hard-plastic toothbrush or something similar. This only works on tooling leather, not full grain. For that, you will just need to make certain the "hairs" from the wrong side are cut away.

If you didn't manage to keep the razor knife or scissors at a 90° angle when cutting out the pieces, now would be a good time to try to straighten up the edges if possible, with a razor knife. Straighten edges before smoothing.

Get rid of the strings by snipping away with a small pair of scissors, or trim with a razor knife. Skived pieces tend to have more strings than non-skived pieces.

Rub the tool along the edges until smooth.

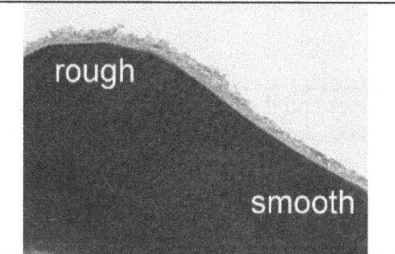

After you have polished and cleaned the edges of the 2 oz tooling leather, you need to add color to the edges. Fiebing's Edge Coat with applicator works fine. Apply from the back, keeping the finished side away from the bottle just in case you slip. If you get any on the finished side, wipe away immediately with a towel, pushing to the edge so you don't make it worse. You can also apply with a small paintbrush for better control.

Blocking Form Tool

While creating the Aussie saddle pattern, the back of the seat proved to be one of the challenges. The biggest problem was wrinkles and folds while I tried to get the leather to conform to the rounded shape. The solution: block the leather first. This is the process of creating a form the desired shape of the leather, wetting the leather, then molding it to that shape and letting it dry. The blocking form tools are made once (for each size saddle you want to make) and reused. As you gain experience, you may not even need the tool.

You will need scraps of about 1/8" thick or thicker (belt quality about 6 to 9 oz) tooling leather or something similar to make these tools. Tandy sells bags of scraps that should provide you with enough small pieces at a fairly reasonable price.

Use stick glue to apply the pattern to the leather. Cut out the pieces. Cut out the center of the top piece and discard.	
Optional but helpful: score along the line for the seat cover piece.	
Remove the paper and then mark the score line with a Sharpie. This will help you position the seat cover piece.	
Glue the seat top piece on top of the tree form piece.	

Glue the seat back piece to the bottom of the tree form piece, wrong sides together.	
Clamp and let the glue dry completely.	
What the finished form piece looks like.	
Depending upon how good your cutting skills are, you may have to carve the back of the seat to get the three layers flush.	Back of seat
Optionally, and depending upon skill level, angle the back of the form downwards.	Back of seat
Check for fit. The form piece should fit snugly into the base. If the form fits too snugly or catches, shave a bit **off of the base** and not the form. Since I have quite a few form tools and make them for different size saddles, I always mark the form pieces with the type and the size. T1 is for the largest of the sizes.	

Pre – Assembly and Prep Work

The following instruction is provided step by step, but not necessarily in the best order for speed. I thought it best to keep the assembly in order. But you may want to proceed with other steps while the glue or leather dries. As you gain experience, you will come up with your own procedure.

Making Keepers

Bridles, breast collars, cinches and stirrups all have keeper straps. The procedure to make them only differs by the keeper type. Fixed keepers are glued to the wrong side of the bottom strap and do not move. Slider keepers are glued into a loop that moves freely along the length of the straps. The keeper straps can be made of leather lace or skiver. The width will vary as will the length. I show the procedure once here rather than bore you over and over again with the same set of instructions.

For a fixed keeper:

Flip the piece so the wrong side faces up. Position the wrong side up with enough on each side to cover the width of the straps. A fixed keeper normally lies below a buckle, which lies under my thumb in this picture. The fixed keeper will go around at least two pieces of leather lace. I like to hold the actual lace parallel and flush behind the buckle strap to make certain I create a large enough loop.	
Place glue on the small side. Press into place on the back side of the buckle strap.	
Spread glue on the other side of the keeper strap and fold over. Trim flush.	

For a slider keeper:

Slider keepers usually go around at least two, more often three strips of leather. Hold all layers of the leather lace flush together with wrong side facing up (even if the backside is finished). Extend one side of the keeper strap further than the width of the lace.	
Fold over flush to the other side. DO NOT glue into place.	
Spread glue on the other side of the keeper strap. Fold over and glue onto the grain side of the other folded keeper strap. Trim flush any excess.	

Making D-Rings from Jump Rings

Less expensive than purchasing D-rings from Rio Rondo is to make them out of jump rings. I believe the Rio Rondo D-rings look better. But for practice saddles, and when funds are short, I use this method.

Position the jump ring with the open part in the middle of the jaw of the pliers. Press firmly. Toothed pliers work better here than smooth jaw pliers.	
What the rings look like after transformation.	

Using the Blocking Form Tool

Tools: Blocking Form Tool, clamps.
Supplies: None.
Materials: Seat cover pattern piece.

Soak the skiver seat cover piece in clean water. Towel off most of the water. Place on the top piece of the blocking tool using the guidelines.	
Place the seat piece in position and push into the hole.	
What the tool looks like from the top.	

Pull and shape the back of the skiver against the form, pulling out the wrinkles. The extra skiver can be folded under the seat.	
Clamp the base to the top and set aside to dry completely.	
Once dry, remove the skiver from the form. It should look something like this from the side.	

Assembling the Tree

Tools: Stylus, scissors, clamps.
Supplies: Glue.
Materials: Tree seat top, tree seat base, tree cantle support, tree aluminum support and seat cover.

Glue the aluminum support piece to the wrong side of the seat top piece.	
Glue the seat base to the seat top and aluminum support pieces.	

Glue the cantle support to the seat base, matching the notches. Let the glue dry. Clamps can be used if needed. This is the completed tree. It still needs to be covered and shaped.	
Spread glue on the seat top piece and on the aluminum wings.	
Place the tree into the depression left by the seat form tool. This is the view from the bottom.	
Start with the front of the tree. Spread glue on the front flap	

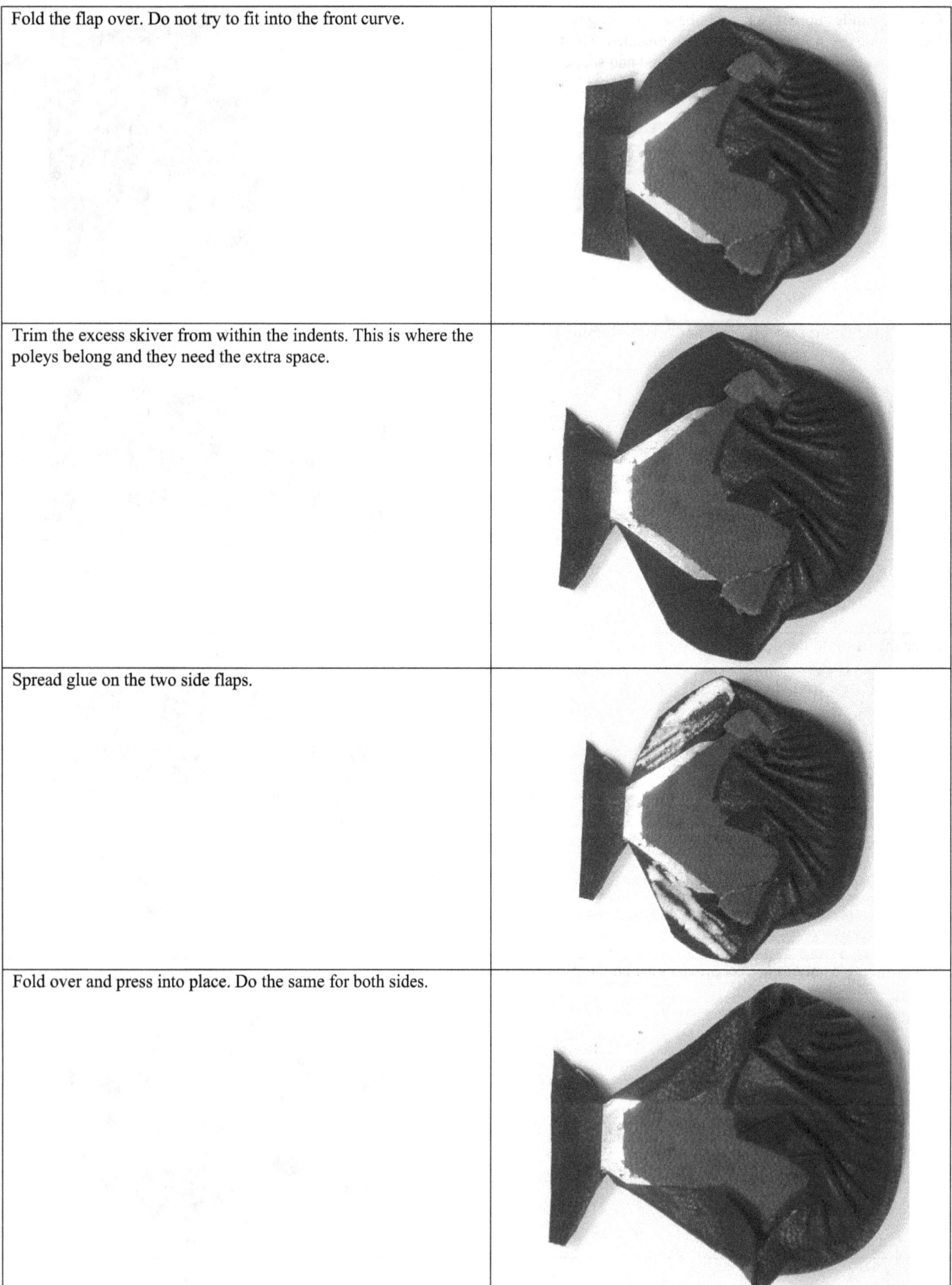

Fold the flap over. Do not try to fit into the front curve.

Trim the excess skiver from within the indents. This is where the poleys belong and they need the extra space.

Spread glue on the two side flaps.

Fold over and press into place. Do the same for both sides.

Pull back the folded skiver from the back of the seat.	
Spread glue along the cantle support piece.	
I like to use white glue because it gives me extra time to work the leather before it sets. This is one situation where a little extra working time comes in handy. Pull the skiver towards the center of the tree working out the wrinkles and working the excess skiver into small rolls. Get the back of the saddle as flat as possible.	
I often use a stylus to help work the leather into position. Here I have flattened out the rolls and worked out as many wrinkles as possible.	

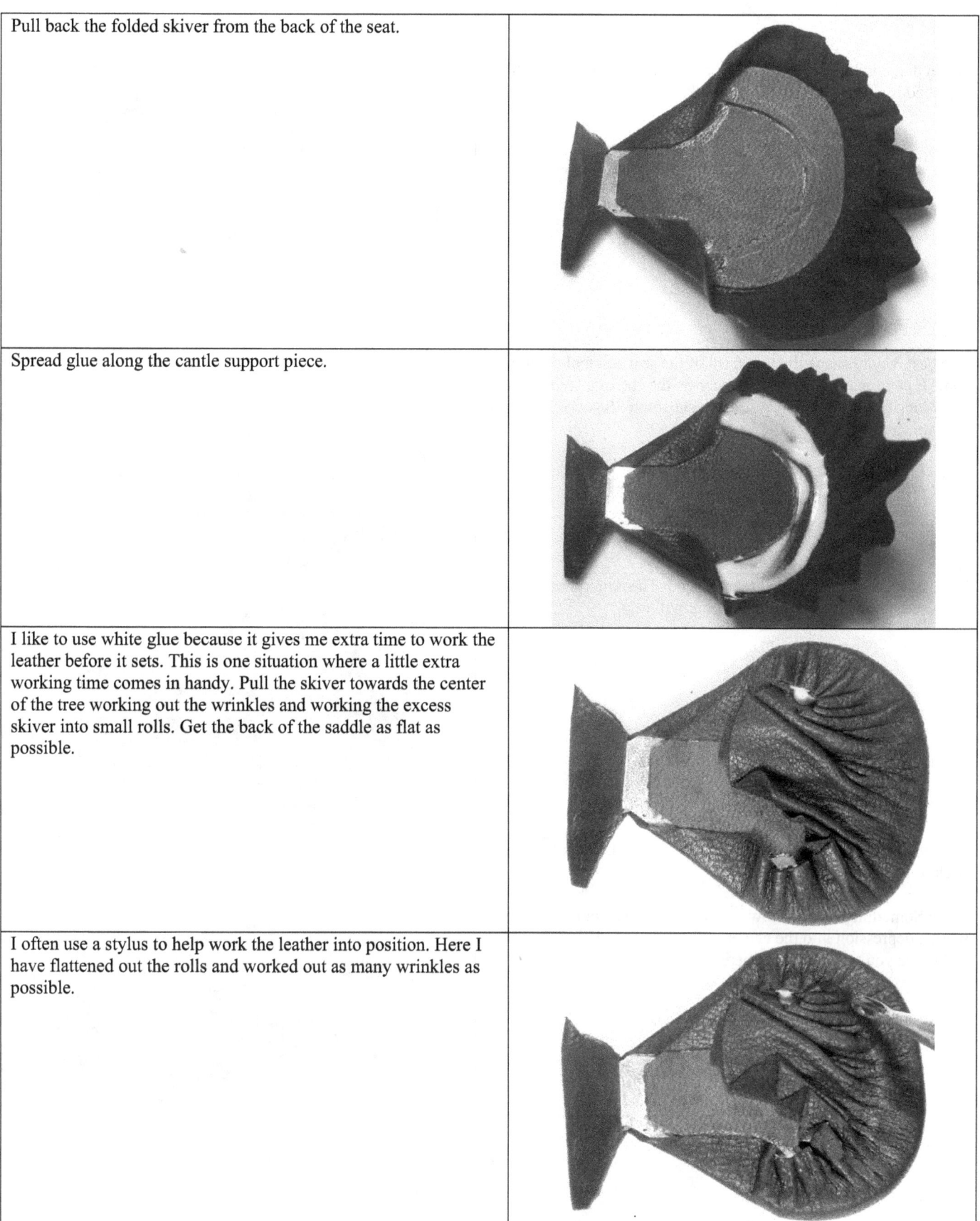

Let the glue dry.

When dry, trim the excess skiver that lies in the center of the tree. It just adds extra and unnecessary bulk.

The best way to shape the tree front of the seat assembly is to use some type of narrow, round tool such as the end of a stylus that has a long point. An awl is a suitable substitute. Just take care not to poke yourself.

Place the tool in the center of the seat assembly as shown.

Press down on the wings to create a bend. The larger diameter tool used, the wider the gullet opening will be.

Push the back of the seat up towards the pommel. Work the seat back to form a depression.

Note: Something I can't show in a book is to use your thumb to work a depression into the center of the seat area. This is a technique you will have to experiment with.

Use the stylus tool to press in the extra leather in the front of the tree.	
What the finished and covered saddletree looks like.	

Piping Detail

Tools: Scissors.
Supplies: Glue.
Materials: Piping detail pieces.

Work the long edge of the skiver and roll as tightly as possible into a log shape.	
Spread glue about 1/8" from the edge.	
Roll the edge of the skiver into the glue.	
Do the same for all three pieces.	

Adding Rear Facing Studs

Tools: Pliers, wire cutters.
Supplies: Glue.
Materials: Straight pins, rear facing pattern piece.

Many Aussie saddles have decorative studs in the rear facing. This detail is easily created with straight pinheads. Depending upon the stitching pattern, use four to eight pins.	
Push the pins in place. Here I have chosen the large oval for the studs rather than the small diamonds, although either would work. Push the pins in about half way.	
Spread a drop of glue on the back of the pinhead. I like to use fast dry gel glue for this detail.	
Push the pins all the way through into position. Wipe away any excess glue. Let the glue dry completely.	

What the piece looks like from the back.	
To keep the pins from flying and potentially hurting someone (or getting lost in the carpet), hold the end of the pin with a pair of pliers. Push down on the leather with the cutters to get as flush a cut as possible. You can discard the rest of the pin or save for buckle tongues.	
What the back looks like with pins removed.	
If you forget to wipe away the extra glue, you will end up with shiny spots.	

Assembling the Panel

Tools: Sewing machine or needle.
Supplies: Matching thread, stylus or stuffing tool.
Materials: Top and bottom panel pieces, extra side stuffing pieces.

Remove the paper from the bottom panel piece, being careful not to tear it.	
Place the top panel piece into position, matching the sides and back to the bottom panel piece.	
Use the paper as a guide to sew the top panel piece to the bottom panel piece around the back edge. Though this can be done by hand, a sewing machine works better. When using the machine, be certain to back stitch over the beginning and ending to lock the stitches into place. For the curved areas, sew straight for as far as possible. Place the needle in the fabric, lift the presser foot and then turn. Repeat until complete. Don't try to force the fabric to turn, as felt tends to stretch out of shape.	
This is what the sewn piece looks like now.	

Remove the guide paper. The stitches should work like perforations in the paper. Some pieces of paper may get stuck under the stitches. Push those loose with the pointed end of the stylus.	
Turn the work right side out. Use the stylus to shape the back of the panel.	
Use the bottom panel paper to guide the gullet stitch. You may find using a bit of stick glue helpful in keeping the paper in position. Match the front of the paper to the panel, making certain the back is centered. The back has a seam at this time and does not exactly match the shape of the paper.	
Stitch along the two center lines the full length of the piece.	

Remove the paper and any little pieces that may get stuck under the stitches.

Push one of the stuffing pieces into the back of the panel.

Stuff the other side and leave the center part flat. Only the back of panel has stuffing. The thickness of the felt puffs out the rest of the panel.

Side view of the stuffed panel piece.

Adding the Rear Facing

Tools: Pliers, wire cutters.
Supplies: Glue.
Materials: Straight pin, assembled panel, rear facing.

Spread glue on the wrong side of the rear panel piece but not on the tail.	
Carefully work the rear facing around the back of the panel with the bottom edge just covering the seam line. The tail on the rear facing should be centered between the gullet seam lines.	
What the panel and rear facing looks like from the side.	
It is more important for the bottom and sides of the rear facing to be glued into place than it is for the top.	

What the assembled piece looks like from the back.	
Now it is time to make the crupper bar. Snip the head off of a straight pin. Use the pliers to bend the pin into a U shape as close to centered as possible. The width of the bend should be equal to the size of leather lace that will be used for the crupper, normally 1/8".	
Slip the pin through the seam in the panel and through the top. The tail should be on the inside of the pin.	
Glue the tail down between the gullet seam lines on the panel.	

Bend the ends of the pin sideways to lock into position. Do the same for both ends.	
The crupper bar pin can be pulled out for tacking up and pushed back in for show time. This is what the finished English and Western panels look like.	

Assembling the Poleys

Tools: None.
Supplies: Glue.
Materials: Front poley fillers, front and rear poley covers, lower flaps, poley panel covers.

Matching the wrong sides, glue the front poley fillers into position, flush to the bottom of the opening.	
With the finished side out, glue the poley covers into position matching bottom, top and all sides.	
Match the rear poley pieces wrong sides together and glue into position.	

What the English version with rear poleys looks like.	
Glue the poley panel covers into position on the wrong side of the lower flap matching the curve to the front and the notch in the back.	
When the glue sets up, shape the front poleys by pressing down and bending the sides forward to produce a slight curve. Shape the back poleys the same way but with less forward curve.	
What the finished English piece looks like.	

Adding the Poley Panels

Tools: None.
Supplies: Glue.
Materials: Front and rear poley panels, upper flaps.

Spread glue on the poley panels as shown. For the front poley panel, you do not want to glue down the edge that covers the stirrup leather opening. For the rear poley panel, spread the glue everywhere. This is actually too much glue, but I just needed to make certain you could see it. Spread all the way to the edge.	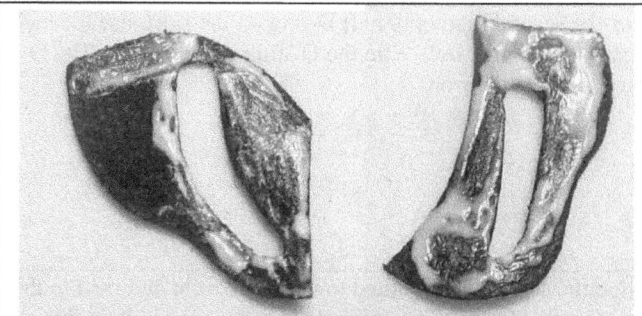
Match the curves and place in position. Wipe away any excess glue.	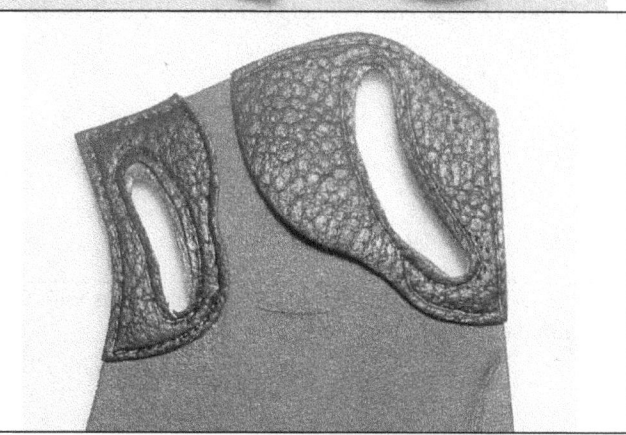

Placing the D-Rings

Tools: Awl.
Supplies: Glue.
Materials: Skirt/flaps for the Western version or skirts for the English version, eight 1/16" D-rings, 1/16" skiver strips. *Note: 1/8" D-rings and 1/8" skived leather lace can be used as a substitute for the front D-rings, if you plan to use the wider lace for martingale or breast collar straps. D-rings larger than 1/8" will look out of scale.*

Glue the D-ring to the D-strap skiver. The finished piece should be no longer than ½". Try for ¼" to reduce the bulk. MAKE 8.	
Use the awl to open the martingale hole at the front.	

Slip the D-strap into position and glue down tab. The D-ring should extend past the front of the flap.	
For the second (and optional) D-ring on the skirt, use the awl again to open the hole. Slip the D-strap into position. The D-rings face towards the front.	
Glue the second strap pulled towards the front and next to the martingale D-ring. This will help to reduce the bulk in this area.	
For the D-rings along the skirt back, open the holes with the awl and slip the D-straps into place.	
Glue into position with the tails of the D-straps pointing towards the notches in the skirt.	
What the finished Western version looks like. Do the same for both sides.	

The English version has smaller skirts, but the procedure is exactly the same. This is the finished piece. Do the same for both sides.	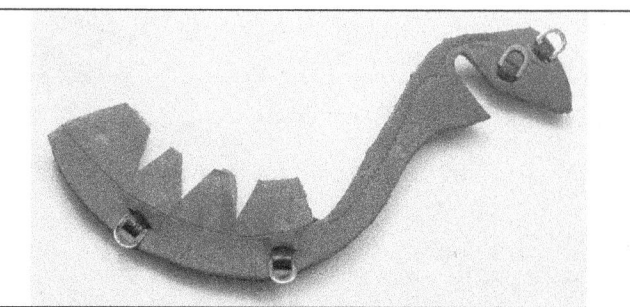

Assembling the Western Stirrups

Tools: Pliers, wire cutters, clamps.
Supplies: Glue.
Materials: Aluminum can strips about ¼" wide and ½" long (optional), stirrups and stirrup liners, 2 straight pins, stirrup treads.

Glue one aluminum strip to the center of the stirrup liner.	
Spread glue on the liner and aluminum leaving one end unglued.	
Place the stirrup on the stirrup liner and clamp into place until the glue dries.	
Cut the tread to fit if necessary, before gluing. Glue the stirrup tread to the center of the stirrup, with the seam on the bottom of the stirrup not the liner.	
On the glued side of the stirrup, push a straight pin through, centered and about 1/8" from the top edge.	
On the unglued side of the stirrup, pull back the outer stirrup piece. Use the awl to make a hole centered and about 1/8" from the top, in the lining piece.	

Push the straight pin through the hole.	
The width of the top of the stirrup (not including the leather) should be about ¼". Use pliers or the stirrup spacer piece to hold the pin at the proper width.	
Bend the straight pin at a sharp 90-degree angle.	
Cut away all but about 1/8" to ¼" of the straight pin.	
Pull the outer stirrup piece back. Center the tail of the straight pin between the outer stirrup and liner. Glue the leather ends together flush.	

Shape the stirrup. Use the tips of the pliers to shape the aluminum into the familiar bell curve. *Wait for the glue to dry before actually shaping the stirrup. Here, I was on a roll taking pictures and didn't wait. The shape of the stirrup is good, but the seam on the cut pin side is coming apart.*	
Wrap the stirrup spacer piece around the pin and glue wrong side together flush at the top. Note: I used contrasting leather for the stirrup spacer piece. Yours should be of a matching color.	
Cut the top nearly flush with the top of the stirrup and trim the sides to create the shape shown.	

English Assembly

Stirrup Bar and Billets

Tools: None.
Supplies: Glue.
Materials: Two 1/4" D-rings, stirrup bar piece, billet piece.

Glue one D-ring to one end of the stirrup bar.	
Check for length before gluing the other Dee. The D-rings should hide in the wells in the lower flap just behind the front poleys.	
Glue into position. The stirrup bar should be slightly curved.	
Spread glue just behind the stirrup bar.	
Center the billet piece and glue just behind the stirrup bar.	
When the glue sets up, pull the billet straps slightly forward.	

Skirts and Upper Flaps

Tools: Scissors.
Supplies: Glue.
Materials: English skirts, seat assembly.

Spread glue along the channel on the seat assembly piece, from the front to where the channel stops before the seat.	
Match the curve of the skirt to the curve in the seat assembly. Press into place.	
Do the same for the other side. Let the glue dry for a bit or the front will come loose when working with the flaps.	
Spread glue on the tabs of the back of the skirt piece. The glue should not extend past the score line.	

Pull the skirt tab around to the backside and bottom of the seat assembly so that the outside edge of the skirt hangs over the back edge of the seat (Top of picture).	
Quickly do the same for the other side. The ends of the tabs should just touch, and the gap should be centered to the back of the seat.	
Use one of the rolled piping pieces, preferably the thinnest roll. Cut off the extra leather, leaving a tube. Discard the tab or save it for buckle keepers.	
Spread glue in the gap between the skirt piece flaps and the seat assembly all the way around the back of the seat. Note: There is too much glue here, but I wanted to make certain it would show up in the picture.	
Tuck the tube of leather into the gap with the seam side facing in. The ends should fit into the small gap where the skirts leave the seat assembly channel. You may need to trim off excess on one end of the leather tube. Wipe away any excess glue.	
What the back of the finished piece looks like.	

Front of Seat Piping

Tools: Scissors.
Supplies: Glue.
Materials: Seat assembly, short front piping piece.

Spread glue on the wrong side tab of the short front piping piece. The glue can extend up the front of the tube.	
Fold the short front piping piece to find the center and then press to the inner and upper edge of the seat assembly pommel area.	
Pull the sides of the short front piping piece down and around the front edge, matching the curved shape of the skirt piece. Trim away the excess in the poley gap in the skirt piece.	
I also trim the front of the piping at an angle that matches the shape of the front of the skirt piece. Do the same for the other side. This is what the finished piece should look like.	

Stirrup Leathers and Flaps

Tools: Clamps.
Supplies: Glue.
Materials: Upper and lower flaps. Skived leather lace 1/8": two 9" lengths.

I find it much easier to add the stirrup leathers at this time instead of at the end of assembly. The stirrup bar rings are difficult to manipulate under the front poley panel. Slip one length of 1/8" leather lace through the stirrup bar D-rings on each side of the lower flap assembly. Pull half way through.	
Slip the ends of the leather lace through the wrong side in the rounded stirrup slit of the upper flap. Make certain that you use the correct piece for the correct side. Pull the lace all the way through.	
Match the poley slot in the upper flap with the poley in the lower flap. Spread glue in the area in front of the front poley. Press the pieces together so that the fronts of both flaps are flush and the front poley fits centered to the poley hole. Clamp if needed while drying.	
Turn the piece around and spread glue along the top edge of the upper flap to rear poley hole. Here I show glue on both pieces but that is not necessary.	
Press the piece into position with the rear poley fitting centered in the rear poley hole. The lower flap will extend past the upper flap.	
Do the same for the other side.	

Seat to Upper Flaps

Tools: Clamps, scissors.
Supplies: Glue.
Materials: Seat assembly, flap assembly.

Turn the seat assembly over. This is the underside of the pommel. Note that there are folds in the piping piece that resemble darts used to make curve reductions in clothing.	
Trim the folds away to reduce the bulk.	
Spread glue in the center of the flap assembly, extending along the top edges of the upper flaps and around the front of the front poley holes.	
Place the seat assembly on top, matching the slits in the skirts with the front poleys. Fit as snugly as possible, even forcing a curve into the center of the flap assembly. The back of the seat should be centered with the tabs of the skirts just covering the top of the rear poleys. This is normal and the coverage will change once the saddle is trained.	
Clamp and let the glue dry.	

Front Piping - Long Piece

Tools: None.
Supplies: Glue.
Materials: Saddle assembly, long front piping piece.

Spread glue on the wrong side tab of the front piping piece.	
Fold in half to find the center and press into the front pommel of the saddle assembly, flush with the short front piping piece. Note: You might find it easier to just glue the center and then let the glue firmly set up before proceeding.	
Pull the end of the long front piping piece along the front curve of the skirt and angle so that the piping disappears in the back of the saddle assembly, at the notch where the lower flaps extend past the upper flaps.	Notch
Do the same for the other side.	

Panel to Flaps

Tools: Clamps.
Supplies: Glue.
Materials: Saddle assembly, panel assembly.

Spread glue on the front inside area of the panel assembly. Note: The craft felt absorbs moisture so glue control can be an issue here. I prefer to use Tacky glue for this part, since it is not as wet as the leather glue.	
Position the panel assembly flush with the front edge and centered to the saddle assembly.	
Continue the side folds on the panel assembly. Glue into position. Do the same for the other side.	
Open the gap between the panel assembly and the saddle assembly and apply glue to the craft felt only. Press into position. The skirt tabs should just cover the inside curve of the rear facing piece.	

Spread glue on the skirt tabs just behind the rear poley.	
Press into position. Clamp if needed. Do the same for the other side.	
Address the small gap between the skirt tabs and the rear facing piece with glue.	
Press into position.	

Stirrup Leathers – Fixed

Buckles on the stirrup leathers add additional bulk that I don't particularly like. With a real Aussie saddle, the stirrup leather buckle would cover the stirrup keeper. The fixed style, though lacking a buckle, actually looks more realistic.

Tools: None.
Supplies: Glue, two English stirrups.
Materials: Saddle assembly.

Make or finish a pair of stirrups. The Rio Rondo stirrups have rough edges that need to be filed down and the entire piece polished. Rio Rondo has a brochure available for free with stirrup purchase that explains how to do this.	
Slip one stirrup onto the front piece of leather lace on the saddle assembly. Be certain to run the lace downwards as shown.	
Glue down about a ¼" tab.	
Position the stirrup correctly. It should extend past the lower flap no shorter than shown. Place the saddle assembly on a body box model and fit so that the bottoms of the stirrups extend even with the bottom of the belly of the model.	

Place a dab of glue on the leather lace where it exits the stirrup bar hole. Press into position.

Place another dab of glue on the leather lace just above the stirrup. Press into position with the edges of the lace flush.

It is correct for the extra stirrup leather to extend past the stirrup as shown.

Position the stirrup keeper just above the stirrup so that the seam is on the inside. Fold over one side without glue. Spread glue on the other side, fold over and press into position. The keeper should be able to slide along the length of the stirrup leather.

Stirrup Leathers – Adjustable

Tools: Scissors.
Supplies: Glue, two 1/8" buckles, two English style stirrups.
Materials: Saddle assembly, stirrup keeper pieces.

Slip one stirrup onto the front piece of leather lace on the saddle assembly. Be certain to run the lace downwards as shown.	
Glue a buckle to the end of the front piece of lace.	
Slip the lower lace into the buckle. The stirrup should extend past the lower flap. In this picture the stirrup leather is adjusted too short.	
Pull all three layers of leather lace flush together and turn over so that the free end of the leather lace is on the top. Position the stirrup keeper behind the stack of lace.	
Fold over one side of the stirrup keeper. Apply glue to the far edge of the other side.	

Wrap around and press into position. The keeper should slide freely along the length of the stirrup leather just below the buckle. Do the same for the other side.	
What the piece should look like, except that I made the stirrup leather too short. I kept them short to keep them in the picture.	

Surcingle

Tools: Awl or stylus.
Supplies: None.
Materials: Saddle assembly, surcingle pattern piece.

Use an awl or stylus to open the surcingle slit in the upper flap on both sides.	
Slide one end of the surcingle into each of the slits. The stirrups should be in front of the surcingle. The straighter edge points towards the front.	

Web Style Girth

Decide how long to make the elastic by fitting to a body box model. The elastic should start and end at the edge of the upper flap. About 5" should do it for Traditional size models. Since the elastic stretches, shorter is better than longer.

Tools: None.
Supplies: Four 1'" buckles, glue.
Materials: Girth pattern pieces, about 5" of ½" elastic.

Glue one girth end piece on each end of the elastic (Top picture shows the back. Bottom picture shows the front). Trim away any excess elastic that might be sticking out at the top.	
Slip buckles onto the tabs almost to the bottom of the split. Try to keep the buckles parallel.	
Fold the leather over and glue into position. A little bit of overlap in the back will help strengthen the bond between the elastic and the leather. If you don't like the bulk, the extra can be trimmed flush with the elastic. Do the same for the other side.	
Front view of the finished side.	
The finished girth.	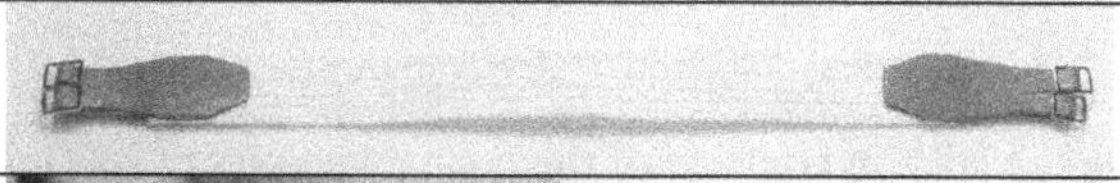
Much like a dressage girth, the buckles should lie closer to the bottom of the lower flap to reduce the bulk under the rider's knees.	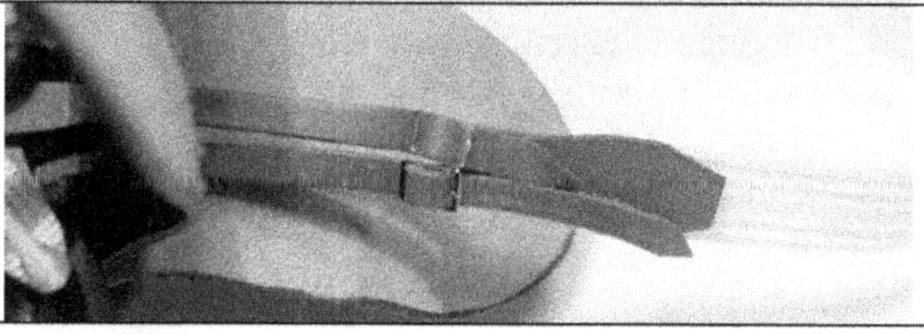

Breastplate

A standard martingale is perfectly acceptable for an English style Aussie saddle. There are actually no rules for what type – if any – of breast collar, breastplate or breast strap to use. Rio Rondo has both a standard martingale and an English breast strap pattern kit available for sale. For the tack pattern collectors, I chose this pattern since it is not available in any other publication. This is a choker style breastplate.

Tools: Awl or stylus
Supplies: Three 1/8" buckles.
Materials: Breastplate pattern piece. Skived leather lace 1/8": two 4" lengths, one 3" length, two 2" lengths and one 1" length.

First make the tie-down loop. Use an awl or stylus to open one of the slits in the center of the breastplate.	
Slip the end of the 1" piece of leather lace into the slit with the grain side up.	
Open the other hole in the center of the breastplate piece and slip the other end of the 1" piece of leather lace as shown.	
Slip the awl or stylus into the loop and turn over. This will help keep the loop open. Glue both ends of the grain side of the leather lace to the wrong side of the breastplate piece. Remove the awl or stylus and turn the piece over.	
Open the slits on one side of the breastplate piece. Starting at the end and working to the center, weave one 4" length through the slits as shown.	
Pull tight with most of the excess lace extending past the end of the piece. Glue down the grain side of the leather lace to the wrong side of the breastplate at the center and also at the end. This is the back of the piece.	
Do the same for the other side. The piece should look like this.	

Glue a buckle to the end of the leather lace.	
Glue the wrong side of one 2" piece of leather lace flush with the lace on the wrong side of the breastplate.	
Buckle the buckle with the free end of leather lace.	
Make one sliding keeper above the buckle and one below the buckle. You can use leather lace. Here I used strips of skiver.	
Slip the free end of the lace into the front pommel D-ring and connect the buckle.	
Glue a 1" piece of skived leather lace to the open area in back between the tie-down loop tabs.	
Use a 3" piece of skived leather lace to make a buckle loop. Glue a buckle on one free end. Make a loop large enough to fit around the girth and glue the other free end to the wrong side of the lace.	
Slip the free end of the center breastplate lace into the buckle.	
Make a keeper out of skiver or leather lace just below the buckle. This is how the loop should fit around the girth when on the model.	

Saddle Blanket – Competition Style

I can't possibly cover all of the different variations available for saddle pad construction. Perhaps I should write a book on the subject? This is one piece that can be different shapes and made of many different materials. Saddle pad construction will be limited by your skills and patience and by how much time you are willing to devote.

The easiest way is to just cut the pattern from a piece of fabric that doesn't fray, like a piece of craft felt. Another idea is to cut two pieces from a fabric that does fray (seam allowances are provided with the pattern), then sew the pieces right side together, leaving a small opening. Turn right sides out, iron flat and either glue or stitch the opening shut.	

For a quilted pad, cut one piece of sheeting about a ½" larger than the finished piece. Use a quilting pencil and draw the pattern on the right side of the cloth.
    Cut a matching piece of batting or craft felt. I use that handy glue stick on the backside of the sheeting to keep the two pieces together while I sew in the quilting lines.
    I used a sewing machine with the smallest stitch possible. On a diagonal, sew in quilting lines about 1/8" apart (or smaller) creating a diamond pattern. Over-sew the edges of the pencil line. Don't worry about back-stitching the ends.	

Trim the piece, cutting away the traced pattern lines. Use bias tape, or cut a 13" piece of sheeting at a diagonal to the grain to make a bias strip. Sew the bias strip, right sides together, as close to the edge of the pad as possible. Curves can be tricky on a machine, but hand sewing takes more time.	
Turn the bias strip right side up and flatten to the back of the saddle pad. Trim away some of the excess to get rid of any lumps. Finally, cut another piece of sheeting fabric and glue to the back of the sewn pattern piece. Don't use too much glue, or the pad may stick to the model. Use just enough to keep the edges down.	

Western Assembly

Stirrup Bar and Billet Flap

Tools: None.
Supplies: Glue, two ¼" D-rings.
Materials: Stirrup bar strap, seat assembly, billet strap.

Glue one D-ring on one end of the stirrup bar strap.	
Check for fit for the second D-ring. The D-ring should be set under the seat and inside the stirrup bar flap in the upper skirt/flap piece. Finger crimp the location and glue the D-ring into position.	
Spread glue about 1/8" wide under the seat between the stirrup bar flaps.	
Press the stirrup bar into position.	

Spread another line of glue about 1/8" wide behind the stirrup bar strap.	
Center the billet strap and glue into position behind the stirrup bar strap.	

Skirts and Upper Flaps

Tools: None.
Supplies: Glue.
Materials: Skirt/upper flap pieces, long piping piece, seat assembly.

Spread glue on the wrong side of the top edge of the skirt/upper flap piece. The glue should not extend into the tail or tabs of the skirt piece.	
Match the curve of the skirt to the curve in the seat assembly. Press into place.	

Do the same for the other side.

Let the glue dry for a bit or the front will come loose when you are working with the skirt tabs.

Spread glue on the tabs of the back of the skirt piece. The glue should not extend past the score line.

Pull the skirt tab around to the backside and bottom of the seat assembly so that the outside edge of the skirt hangs over the back edge of the seat.

Quickly do the same for the other side. The ends of the tabs should just touch, and the gap should be centered to the back of the seat.

Use the rolled piping piece you made into a tube.

Spread glue in the gap between the skirt piece flaps and the seat assembly all the way around the back of the seat.

Tuck the tube of piping leather into the gap with the seam side facing in. The ends should fit into the small gap where the skirts leave the seat assembly channel. You may need to trim off excess on one end of the leather tube. Wipe away any excess glue before it dries.

Stirrup Fenders – Fixed

Note: I skipped to this step while the skirt pieces dried, before gluing down the skirt tabs.
Tools: None.
Supplies: Glue.
Materials: Stirrups, stirrup fenders, stirrup keepers, saddle assembly.

Pull the stirrup bar flap up and locate the stirrup bar D-ring.	
Run the right stirrup fender strap through the slot and through the Dee.	
Pull the end back through the stirrup bar slot. Do the same for the other side.	
Glue the end of the top fender leather to the back of the fender flush or slightly inward.	

Slip the other end through the stirrup. Check for correct length; stirrup bottom should end at the bottom of the model's belly. Glue the free end to the back of the stirrup leather.	
With the back of the fender facing you, place the stirrup keeper wrong side up around the straps of the stirrup fender.	
Fold over one side. Place glue on the other side and fold over.	
What the piece looks like when finished. Center the stirrup fender strap in the center of the stirrup.	

Stirrup Fenders – Adjustable

I don't use buckles for the stirrup fenders because it adds bulk that I find annoying. But some people prefer adjustable stirrups.
Tools: None.
Supplies: One 1/8" buckle.
Materials: Stirrups, stirrup fenders, stirrup keepers, saddle assembly.

Pull the stirrup bar flap up and locate the stirrup bar D-ring.	
Run the right stirrup fender strap through the stirrup bar slot and through the Dee.	
Pull the end back through the stirrup bar slot. Do the same for the other side.	
Glue a buckle to the other end of the stirrup fender with a ¼" tab.	

Slip the end through the stirrup and buckle.	
With the back of the fender facing up, place the stirrup keeper wrong side up, around the straps of the stirrup fender.	
Fold over one side. Place glue on the other side and fold over.	
What the piece looks like when finished. Center the stirrup fender strap in the center of the stirrup. Make certain that the staps move freely inside the sliding keeper.	

Upper Flaps to Lower Flaps

Tools: Clamps.
Supplies: Glue.
Materials: Lower flaps, saddle assembly.

Spread glue in the area ahead of the front poleys.	
Slip the poleys into the poley slots of the saddle assembly. The fit should be snug with the poleys in the center of the poley slot. Clamp if needed.	
Open a gap between the lower flaps and the saddle assembly. Spread glue in the gap and press the pieces together. Make certain the pieces are centered and even on both sides.	
The lower flap should extend past the upper flap. Clamp and let the glue to dry.	

Horn

Tools: Wire cutters, pliers.
Supplies: One 1 ½" straight pin or 1 ½" length of 20-gauge wire.
Materials: Horn pattern pieces, horn cover piece.

Snip the head off of the straight pin.	
Using the round nose pliers, make an eye in the end from which the pinhead was removed.	
Spread glue on the horn and small tab of the horn piece.	
Place the eye in the pin centered in the horn area extending down parallel to the side of the horn piece.	

Page 65

Glue the horn cover over the eye in the pin flush around the front curve with the open tab in the center back.	
Fold over the small tab.	
Spread glue on the long tab and part of the short tab that extends past the pin.	
Wrap the long tab around the pin. This creates the neck of the horn.	
Using a pair of toothless pliers or a smooth area of toothed pliers, bend the eye of the pin towards the seam in the horn neck.	

Horn to Seat

Tools: Awl, pliers, wire cutters.
Supplies: Glue.
Materials: Horn assembly, saddle assembly, and horn support pattern piece.

Dry fit the horn support on the pommel. Use an awl to make a hole in the center of the horn support.	
The hole needs to go through the aluminum support in the tree.	
Push the horn into the hole to dry fit.	
If you think the horn neck is too tall, it is okay to cut some of it away.	

Spread a little bit of glue around the bottom of the horn to help secure it to the pommel.

From the underside, bend the pin as close to the bottom of the seat assembly as you can. Trim away some of the excess, leaving just enough to anchor the horn into place on the underside of the pommel.

Spread glue on the wrong side of the horn support. Wrap around the horn and press into place. There should not be any gap in the front of the support.

Glue the tabs to the front and underside of the pommel, wrapping them smoothly.

Piping – Short Piece

Tools: None.
Supplies: Glue.
Materials: Seat assembly, short front piping piece.

Spread glue on the wrong side tab of the short front piping piece. The glue can extend up the front of the tube.	
Fold the short front piping piece to find the center and then press to the inner and upper edge of the seat assembly pommel area.	
Pull the sides of the short front piping piece down and around the front edge, matching the curved shape of the skirt piece.	
Trim away the excess inside the poley gap in the skirt/flap piece. Do the same for the other side. This is what the finished piece should look like up close.	

Piping - Long Piece

Tools: None.
Supplies: Glue.
Materials: Saddle assembly, long front piping piece.

Spread glue on the wrong side tab of the front piping piece. Fold in half to find the center.	
Press into the front pommel of the saddle assembly flush with the short front piping piece.	
Pull the end of the long front piping piece along the front curve of the skirt. Angle it so that the piping disappears in the back of the saddle assembly at the notch where the lower flaps extend past the upper flaps.	
Do the same for the other side.	

Rear Rigging Rings

Tools: None.
Supplies: Two 6mm jump rings, glue.
Materials: Rear rigging flaps. Skived leather lace 1/8": one 2" length.

Glue the 1/8" leather lace to the 6mm jump ring with a ¼" tab. Trim.	
Repeat. Make two tabs on the one jump ring. Do the same for the other jump ring.	
Glue one of the lace tabs to the tab area of the rear-rigging flap with the jump ring centered in the round area. Trim any excess.	
Spread glue on both the grain and wrong sides of the tab on the rear-rigging flap.	
Slip between the upper and lower flaps in the corner just below the seat.	
Glue the other tab to the back of the skirt flap. Do the same to the other side.	

Front Cinch

Tools: Wire cutters, pliers
Supplies: Two 9mm jump rings, two 1/8" D-rings (or two 5mm jump rings made into D-rings), two silver eye pins (or two 2" pieces of 22-gauge silver wire made into eye pins).
Materials: Front cinch piece, front billet keepers, front cinch lining. Skived leather lace 1/8": one 1" length.

Slip the eye pins onto the 9mm jump rings and close the ends. Finished pieces should look like this. Note: if using wire instead of eye pins, use rounded nose pliers to make the loop and then slip onto the 9mm jump rings.	
Slip the wire through the wrong side of the slit in the front cinch piece.	
Tuck the free end of the cinch piece through the cinch ring.	
Fold over. Work the eye of the eye pin into the slit and then glue the tab into position, wrong side to wrong side.	
Glue the 1" piece of leather lace onto the flat side of the 1/8" D-ring.	
Position over the cinch piece to get an idea of proper length. Slip the remaining 1/8" D-ring onto the free end of the lace. Glue over the flat part of the D-ring.	
The finished strap should allow the D-rings to extend just over the edge of the front cinch.	

Fold the front cinch in half to find the exact center. Spread a thin strip of glue about 1/8" wide.	
Place the D-strap on the glue and press into position. Trim the extra cinch ring wire, keeping it just long enough to extend about 1/16" past the end of the ring.	
Spread glue along the wrong side of the lining piece. Match the shape of the front cinch lining piece and press into position. There may be extra lining extending past the cinch rings that needs to be trimmed.	

Rear Cinch

Tools: Pliers, wire cutters.
Supplies: Two 9mm jump rings, one 1/8" D-ring (or one 5mm jump ring made into a Dee), two silver eye pins (or two 2" pieces of 22-gauge silver wire made into eye pins).
Materials: Rear cinch piece, two rear billet keepers, rear cinch lining. Skived leather lace 1/8": one 1" length.

Slip the eye pins onto the 9mm jump rings and close the ends. Finished pieces should look like this. Note: if using wire instead of eye pins, use rounded nose pliers to make the loop and then slip onto the 9mm jump rings.	
Slip the wire through the wrong side of the slit in the front cinch piece.	
Tuck the free end of the cinch piece through the jump ring.	
Fold over. Work the eye of the eye pin into the slit and then glue the tab into position wrong side to wrong side.	

Trim the cinch tongue wire so that it extends 1/16" to 1/8" past the end of the ring.	
I use the rear billets as a guide to make certain I don't glue the front billet keepers on too tightly	
Position one of the front billet keepers just below the cinch ring around both the rear billet piece and the front cinch. Make a fixed keeper on each side.	
Glue the 1" piece of leather lace onto the flat side of the 1/8" D-ring.	
Fold the front cinch in half to find the exact center. Spread a thin strip of glue about 1/8" wide.	
Place the D-strap on the glue and press into position.	
Trim the excess leather lace.	
Spread glue along the wrong side of the piece.	
Match the shape of the front cinch lining piece and press into position. There may be extra lining extending past the cinch rings that needs to be trimmed.	

Front to Rear Cinch Connector Strap

Tools: Scissors.
Supplies: Glue, one 5mm jump ring, one 1/8" buckle.
Materials: Skived leather lace 1/8": one 3" length.

Glue buckle to leather lace.	
Slip the buckle strap through the jump ring.	
Insert through Dee in front cinch.	
Fold over and slip through the center jump ring again.	
Slip through the D-ring in the rear cinch.	
Buckle and slip the extra through the center ring again. The center ring should have three layers of leather lace running through it.	

Surcingle

Tools: Awl or stylus.
Supplies: None.
Materials: Saddle assembly, surcingle strap.

Use an awl or stylus to open the surcingle slit in the upper flap on both sides.	
Slide one end of the surcingle into each of the slits. The stirrups should be in the front of the surcingle. The straighter edge points towards the front.	

Panel to Flaps

Tools: Clamps.
Supplies: Glue.
Materials: Saddle assembly, panel assembly.

Spread glue on the front inside area of the panel assembly. Note: The craft felt absorbs moisture so glue control can be an issue here. I prefer to use Tacky glue for this part since it is not as wet as the leather glue.	
Position the panel assembly flush with the front edge and centered to the saddle assembly. Continue the side folds on the panel assembly. Glue into position. Do the same for the other side.	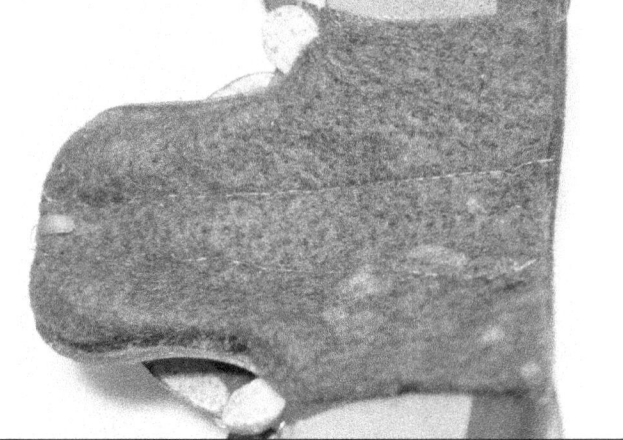

Open the gap between the panel assembly and the saddle assembly and apply glue to the craft felt only. Press into position. The skirt tabs should just cover the inside curve of the rear-facing piece.	
Spread glue on the skirt tabs just behind the rear poley.	
Press into position. Clamp if needed. Do the same for the other side.	
Address the small gap between the skirt tabs and the rear-facing piece with glue. Press into position.	

Rear Billets

Tools: None.
Supplies: Glue
Materials: Rear billets, two rear billet keepers.

Fold the rear billet in half with the wrong side facing out.	
Place the rear billet keeper behind the pointed end of the rear billet with the wrong side facing forward. This will help keep the loop the right size.	
Apply glue to both sides of the rear billet keeper.	
Fold over one edge of the keeper so that it is flush with the side of the rear billet. Press into place.	
Fold over the other side of the keeper pulling it somewhat snug. Clip away any extra that extends past the edge of the side of the billet.	

Clamp and let dry.	
Do the same for the other billet. Pull the pointed end out of the loop. This is what the piece should look like.	
To install the billet, slip through the rear cinch ring with the wrong side up.	
Fold the loop end up.	
Slip the pointed end into the loop through the top.	
Pull tight. Do the same for the other side.	

Page 79

Cinch Billet Adapter

Tools: None.
Supplies: Four 1/8" buckles.
Materials: Skived leather lace 1/8": two 1" to 2" lengths. Note: Make this piece longer or shorter to better fit your model horse.

Glue one 1/8" buckle to each end of the leather lace.	
Make a hole in the exact center of the leather lace. You can make a series of holes for additional adjustments.	
Tuck the adapter through the cinch ring and place the tongue into the hole.	
Do the same for the other side. Buckle to the billets on the saddle as shown. This strap should be 1" shorter for this particular model horse.	

Roper Style Breast Collar

Tools: Scissors and clamps.
Supplies: Two 6mm jump rings made into D-rings, two 5mm jump rings, one 9mm jump ring, three 1/8" buckles, and glue.
Materials: Pattern pieces for breast collar center, breast collar and breast collar lining. Skived leather lace 1/8": two 4" lengths, three 1 ½" lengths.

Glue the 6mm D-rings to the smaller end of the breast collar piece. Do the same for both sides.	
Glue the larger end of the breast collar piece to the 9mm jump ring.	
Do the same for the other side. The piece should curve upwards as shown if properly assembled.	
Spread glue to the wrong side of the breast collar piece.	
Match the breast collar lining to the breast collar piece and glue into position. Do the same for the other side.	
Glue one 1 ½" piece of leather lace to the center bottom of the 9mm jump ring. Keep the folded tab small, about ¼" or so.	
Spread glue onto the RIGHT side of the breast collar centerpiece tab.	

Position the round part of the breast collar centerpiece to the 9mm jump ring. Glue the tab part of the centerpiece to the leather lace.	
Glue one 1/8" buckle to the piece of 1 ½" leather lace.	
Buckle to the breast collar center strap. Make a fixed buckle keeper below the buckle.	
Slip the free end of the leather lace through the center jump ring on the front cinch. Test for correct length on the model, then glue into position. Unbuckle the buckle to complete the assembly. Note: It is more realistic to use a D-ring and a hook for the attachment to the center cinch ring. But I find that the hooks are difficult to attach during live shows (time constraints), and can cause scratches to the model horse paint, between the front legs.	
Glue one buckle to one end of each of the 4" lengths of leather lace.	
Tuck the free end of the leather, right side up, through the 6mm D-ring and fasten the buckle.	
Make one sliding keeper above the buckle and one below the buckle. Do the same for the other side.	
The breast collar can be attached to the pommel D-rings rings in the saddle. An alternative is to slip the strap around the billets and girth of the saddle. Fit to the model and then trim the excess at the 6mm D-ring.	
Buckle the center strap to the cinch strap. Fit to the model and trim the excess.	

Page 82

Saddle Blanket – Western

Tools: Scissors.
Supplies: Tacky glue.
Materials: Saddle blanket fabric, side decorations and blanket center piece.

Glue the sidepieces to the craft felt, flush with the bottom edge and centered.	
Make a slit in the center front of the fabric. This allows extra room for the withers, especially high withered models. But it is not always necessary. Test how the blanket lies on the model you wish to use.	
Place glue around the slit. It might be easier to apply the glue to the blanket centerpiece; that is a personal choice.	
Glue the blanket centerpiece to the center front over the slit. There should be a gap appropriate to the amount of room needed for the model horse withers. This picture shows the wrong side up.	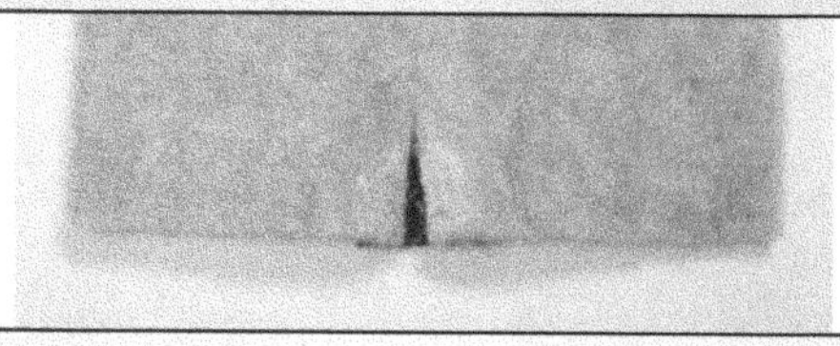
What the finished blanket looks like.	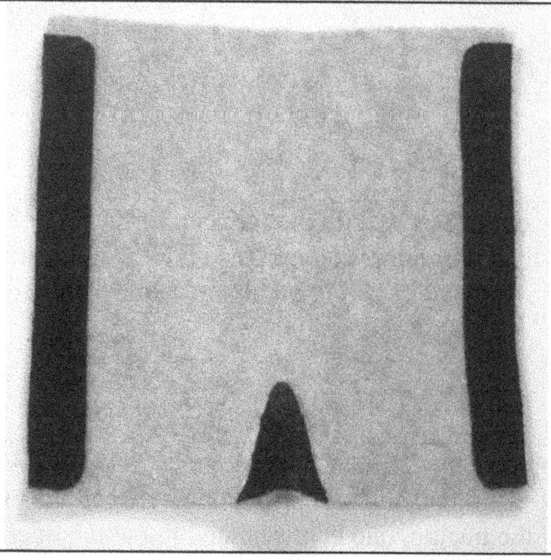

Barco Bridle

Tools: None.
Supplies: Glue, two 7mm jump rings or Cast Snaffle bit, two 6mm jump rings, 4 buckles to match the width of the leather lace.
Materials: Skived leather lace 1/8": two 8" lengths (cheekstraps), one 3 ½" length (browband), one 3" length (crown piece), one 16" length (English reins) and one 4" length or scrap pieces (keepers). Skived leather lace 1/16": one 7 ½" length (throatlatch).

Glue one 6mm jump ring to one side of the 3" strip of lace.	
With the 3" strip, dry fit for the other jump ring. The rings should drape straight across the crown and lie below the ear sockets on both sides of the head. Glue the other 6mm jump ring to the opposite end of the 3" strip. This completes the crown piece. Note: Mark the back of the crown piece, as it may be difficult to distinguish from the browband until the bridle is fully assembled.	
Glue the 3 ½" strip to the 6mm jump ring on one side of the crown piece.	
Dry fit the browband so that the jump ring lies below the ear socket and snug across the forehead. Glue the free end of the browband to the other 6mm jump ring.	
Glue one buckle to one end of each 8" strip of lace for the cheekstraps.	
Take the cheek pieces, slip the free end through the bit, then through the 6mm jump ring and then buckle it, creating a loop with the wrong sides facing in. Do the same for the other side.	1. Bit 2. Ring 3. Buckle
Make three keepers for each cheek piece; one fixed keeper below the buckle and two slider keepers, one between the 6mm jump ring and the buckle, one between the bit and the buckle that wraps around all the layers of lace. Do the same for the other side.	Stationary Slider Slider

Glue the 7 ½" piece of lace to the right side 6mm jump ring between the cheek piece and the crown piece.	
Check for correct length on the model. Trim so that the throatlatch extends just to the 6mm jump ring on the opposite side, perhaps ¼" shorter.	
Glue the trimmed piece to the 6mm jump ring on the other side between the crown piece and the cheek piece.	
Glue a buckle to the right side (longer) throatlatch piece.	
On the buckle side of the throatlatch piece, make one stationary keeper just below the buckle and one slider keeper.	
For the reins, glue one buckle to the end of the 16" length of lace. Buckle with the other end of the lace leaving about a 3/8" tab. Make one stationary keeper next to the buckle. Trim the rein to a pointed end.	
Keep the reins buckled. Fold in half with the buckle as the center.	
Cut in half, making two reins.	

Glue the cut ends of the reins to the bit.	
Fit the bridle to the model using sticky wax to hold the bit into position. Trim away the excess lace that extends over the bits. Trim the throatlatch piece to the throat.	

Halter Bridle

This is one of the most complicated bridles I have ever made, about equal to a double bridle. It has more hardware than any other bridle I have made. The lower part resembles a halter while the upper part resembles a Western headstall. The entire bridle could be made from 1/16" leather lace. I always suggest starting with 1/8" and then slim down to 1/16" as you gain experience. But there are parts on this bridle that require 1/16" lace, such as the throatlatch strap and the bridle hook straps, so I am using a combination of both sizes. Because of the complexity of this bridle, I cut the lace as needed rather than try to keep track of all the small pieces. I always tend to cut the strips longer than needed, as it is always easier to cut away excess than it is to rip the thing apart and start over again. I use the trimmed excess for buckle keepers.
Tools: Pliers, wire cutters, scissors.
Supplies: Glue, two 1/8" (Traditional scale) halter rings, two 5mm jump rings, two 7mm jump rings or cast snaffle bit, four 1/8" buckles, three 1/16" buckles, three 5mm D-rings, two 3mm D-rings, three hooks, 1" length of 24 gauge wire or one 5mm jump ring.
Materials: 10-12" of braided floss or 1/16" leather lace for lead rope. Skived leather lace: 1/8": one 3 ½" length (browband), one 6" length (headstall), two 3" lengths (cheek pieces), one 2" length (noseband), one 3" length (chin strap), one 2" length (chin/throatlatch connector), one 16" length (English reins). Skived leather lace 1/16": one 5" (throatlatch), two 3" (bit hook straps), one 3" length or scraps for buckle keepers.

Glue a halter ring to each side of the noseband strap. Make certain you have a right and left side as shown (round ends pointing in the same direction).	
Make certain the band fits snugly, with the halter rings laying centered to the lips, in line just behind the ear socket, above the lip and below the cheeks.	
Glue one buckle to the chinstrap.	

Slip the chinstrap through the back slits in the halter ring, grain side towards the wrong side of the noseband.	Suede side / Grain side
Buckle the chin strap, leaving about a 1" tail.	
Glue one cheek piece to the remaining open slot in the halter ring. Do the same for the other side.	
For best results, fit to the model by folding over and finger crimping the buckle location. For this bridle the buckle should be located just above or equal to the eye on both sides. Any higher and it will interfere with the browband. Any lower and the bit straps will be too short.	
Slip a 1/8" D-ring on the strap and then place a buckle on the finger crimp location. The D-ring is both the buckle keeper and the bit strap hook connector.	
Glue the buckle and D-ring into position. Keep the flat part of the D-ring just below the end of the buckle. For added strength, glue down a tab flush with the end of the halter ring tab. Trim away the excess.	Trim

Buckle the headstall piece to each of the cheekstrap buckles. The tails should extend all the way to the halter rings on both sides. You can adjust and trim to correct length at the final fitting. Leave them long for now.	
Glue a buckle to one end of the throatlatch piece.	
Place the browband piece wrong side up with the extra pointing away from the bridle. Lay the headstall over the browband grain side up, front up. Lay the throatlatch piece next to the headstall.	
Wrap the long end of the browband piece towards the front of the bridle and fold around the headstall and the throatlatch. Glue down to create a snug pocket in the browband in which the throatlatch and headstall can slip through freely.	
For correct browband fit, place the bridle on the model and finger crimp the correct location on the other side. The browband should be snug, but should not pull the headstall piece into the ear (Wrong as shown. I had trouble getting the leather to cooperate for the picture).	
Remove the bridle from the model and create a pocket for the headstall and throatlatch piece to move through freely. Trim away any excess from the back of the browband.	

Open one of the 5mm jump rings and slip into the center of the other 5mm jump ring. Keep the ring open.	
Slip all three layers of chinstrap leather through the open end of the jump ring. Close the jump ring.	
Glue the chin/throatlatch connector strap to the loose jump ring.	
Fold the other end of the chin/throatlatch strap over and create a loop. You may want to fit this to the model so that the strap isn't too long. Trim away any excess.	
Pull the buckle end of the throatlatch piece through the chin/throatlatch connector strap loop.	

Make a fixed keeper on the throatlatch piece.	
Attach one 1/16" D-ring to a short hook. If using the Rio Rondo D-rings, you will have to use wire cutters to make an opening in the flat end of the D-ring.	
Slip the bit hook strap through the D-ring and pull through about one third of the length.	
Slip both ends of the lace through a 1/16" buckle. Pull through to within 1/8" of the D-ring.	
Pull the short end through the buckle and then the long end on top of the short end. Position the buckle before pulling tight. (This strap is intentionally long because it is easier to do this with a long piece than a short piece).	
Pull tight. Do the same for the other bit hook strap.	
For proper fit, place the bridle on the model. Use sticky wax to hold the 9mm (bit) jump ring into position. Attach the hook end to the D-ring on the cheekstrap. Slip the free end of the bit hook strap through the front of the bit and wrap around. The fit should be snug but not tight enough to dislodge the bit. Finger crimp the correct location.	

Glue the bit ring into position. Trim the excess from the bit ring tab. Keep the buckle strap as long as possible, but trim away any hanging behind the bit ring.	
Do the same for the other side. This is what the piece looks like, front and back.	
For the English style reins, glue one buckle to one end of the 16" leather lace. Buckle the other end of the leather lace with about a 3/8" tab. Make a fixed keeper just below the buckle. Trim the blunt end to a point.	Pointed end
Fold the piece in half with the buckle as the exact center.	
Cut in half on the loop end.	
Glue one end of each rein to each of the bits. Make certain the right side of the rein and the right side of the bit connector straps face the same direction.	
Hook the bit connector strap onto the D-ring on the cheekstrap and crimp it closed. Do the same for the other side.	Crimp closed

For the lead rope, I braided three strands of twisted floss (the twisted type), two strands of black and one of white. I start with a knot and end by wrapping the center strand around the other two and slipping through the loop. This makes a nice small knot on the tassel end.	
On the small knot end, I trim to tassel size and then pull or untwist the thread with an awl to create a used tassel look.	
Attach a long hook to a 1/8" D-ring. If using the Rio Rondo D-rings, you will have to create an opening in the D-ring with a pair of wire cutters. Slide the hook in and close the D-ring.	
Spread glue around the large knot end to keep the strands together. Let the glue just set up a bit. Then trim the knot.	
Slip the end into the D-ring and fold over about ¼". Use more glue if needed to get the ends to stick.	
Use 22-gauge silver wire (about 1" or whatever length you find easy to work with) and wrap around the folded end at least two times. You can use a 5mm jump ring instead of wire. Make certain it wraps around.	
Trim away the excess wire as flush to the rope as possible. Use pliers to push the ends into the rope.	

Attach the hook to the halter ring. Crimp closed if desired.	
Pull the rope around the horse's neck loosely. Put the loose end over the top of the hook end.	
Make a loop and then insert the tassel end into the loop. The tassel should point downwards.	
Pull snug but not too tight. The lead rope should be short so the horse won't trip on it while running the trail. It should be loose enough around the neck so that it does NOT act like a tie-down.	

Crupper Style 1

This pattern is easier than the Crupper 1 pattern. Either pattern will work, but this is more Australian in design.
Tools: Pliers.
Supplies: Two 1/8" buckles.
Materials: Skived leather lace 1/8": two 2 1/2" lengths (rump strap, tail buckle strap) and one 5" length (crupper bar strap).

Glue one buckle to each end of one of the 2 ½" lengths of lace.	
Buckle one end of the 5" length of leather leaving about a ½" tail.	
Fold the 5" length around and glue a ¼" tab of the wrong side to the wrong side of the rump strap, just behind the buckle. Do not glue down the buckle itself.	
Glue the remaining 2 ½" length grain side to the back of the buckle, flush with the end of the crupper bar strap end. Angle it slightly to the side, taking advantage of the natural curve in the lace. This strap can be on either side of the tail; the left side is the most common.	
Buckle the tail straps together. Make one sliding keeper on the crupper bar strap between the crupper bar loop and the buckle. Make one fixed keeper just below the tail strap buckle. Trim the lace ends to points.	
Unbuckle the crupper bar strap and insert into the crupper bar on the saddle.	
Slip back through the sliding keeper and then buckle.	
Unbuckle the tail strap. Wrap around the tail and then buckle. This is what the finished piece looks like.	

Crupper Style 2

Tools: Razor knife or scissors, pliers.
Supplies: Two 16" buckles, one 1/8" buckle.
Materials: Skived leather lace 1/8": one 4" length (crupper bar strap), one 2 ½" length (rump strap), one 1 ¼" length (under-tail strap cover). Skived leather lace 1/16": one 2 ½" length (under-tail strap) or length needed to fit the model.

Glue a buckle to the end of the 2 ½" rump strap. With the 4" crupper bar strap, spread glue on the last ¼" or so of the wrong side.	
Press to the wrong side of the rump strap just behind the buckle. Do not glue the buckle down to this strap.	
Buckle the crupper bar strap to the rump strap buckle.	
Using either a razor knife or scissors, split about 1" of the free end of the rump strap.	
Glue a buckle to both ends of the 1/16" under-tail strap.	
Fold both the under-tail strap and the tail strap in half to find exact center. Here I've marked center with a pen for illustration purposes.	
Trim the ends of the under-tail strap cover to points.	
Glue the grain side of the cover to the wrong side of the tail strap matching centers.	
Make one sliding keeper between the buckle and the crupper bar loop, one fixed keeper below the rump strap buckle and two fixed keepers for each of the tail strap buckles.	

Unbuckle the crupper bar strap. Pull the tail free of the sliding keeper. Use pliers if needed to pull the crupper bar out far enough to make room for the crupper bar strap. Slip the loose end through.	
Slip the tail through the sliding keeper and then buckle.	
Attach the tail strap. Buckle one side, bring under the tail and then buckle the other side. This is what the completed crupper strap looks like.	

Training the Saddle

After assembly the saddle will want to lie flat. It needs to be trained to the model horse shape. Put it on a body box model and place thick rubber bands around to force the leather to curve and lie the way a real saddle does. If you don't have rubber bands, use some type of flat lace (shoelace) and wrap it around the model and saddle. Each saddle will differ depending upon the leather, but most of the time this is an overnight process.

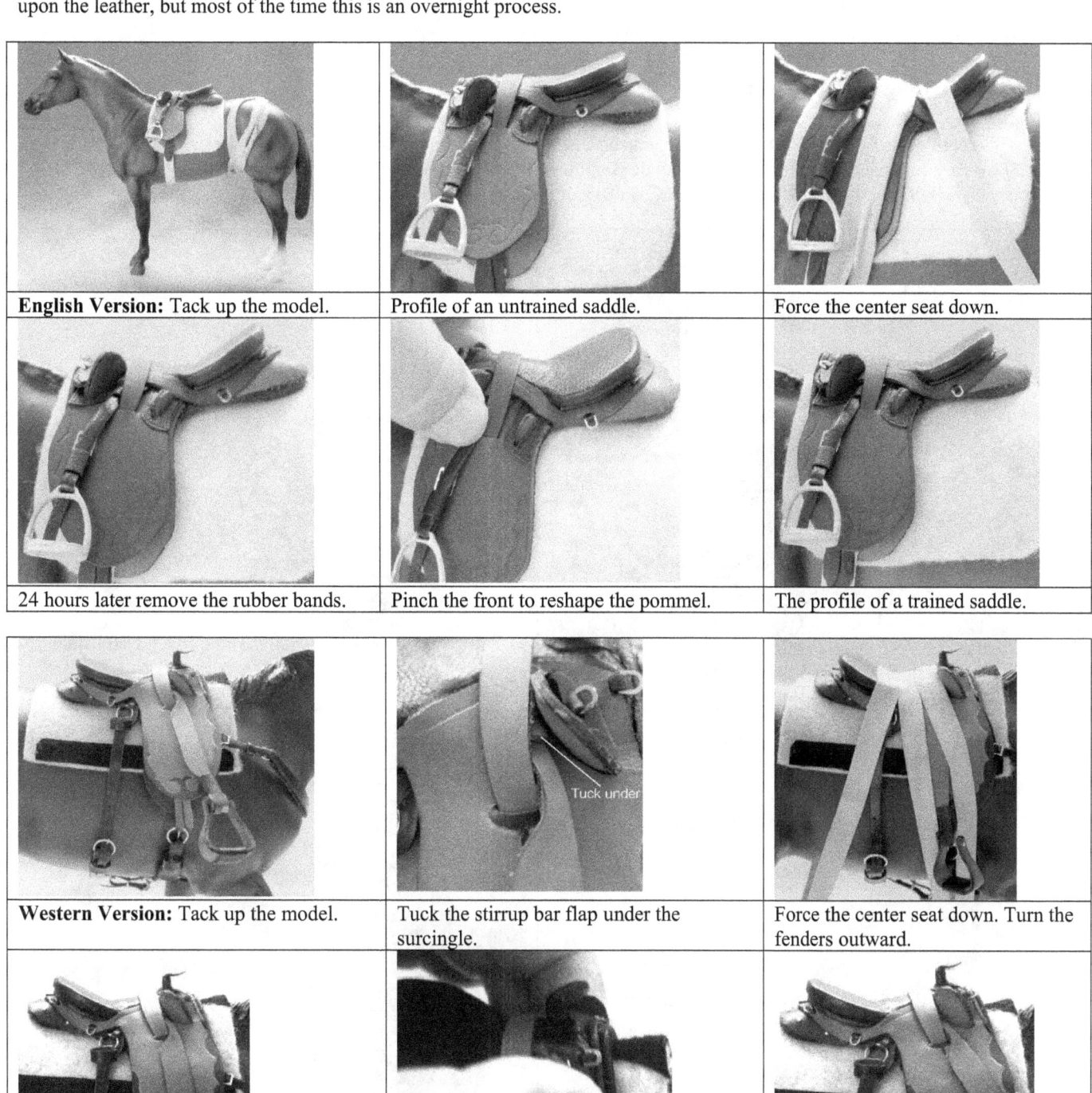

English Version: Tack up the model.	Profile of an untrained saddle.	Force the center seat down.
24 hours later remove the rubber bands.	Pinch the front to reshape the pommel.	The profile of a trained saddle.
Western Version: Tack up the model.	Tuck the stirrup bar flap under the surcingle.	Force the center seat down. Turn the fenders outward.
24 hours later remove the rubber bands.	Pinch the pommel from behind the poleys.	The trained Western version.

Applying a Finish Coat

If you want a nice satin or glossy sheen to the leather, you can finish it with a good acrylic finish coat designed for leather. If you dyed your own leather, this will help with possible bleed through. Bleed through is more of an over time than an overnight process. Don't keep your best models tacked up for long periods of time or the color can transfer from the saddle to the model in any area where the dyed leather comes into contact with the plastic.

If you decide to use any of the antique finishes, a final finish coat is required to keep the antique finish from rubbing off or washing off if you ever need to clean dust off the saddle.

Options and Suggestions

This set of patterns can be called a mix and match set. English doesn't have to be English and Western doesn't have to be Western. Two-toned saddles are quite common. Tooling is optional but is normally found in a U-shaped pattern around the edge of the upper flaps. For other ideas, check out pictures on eBay or Australian Stock saddle distributors.

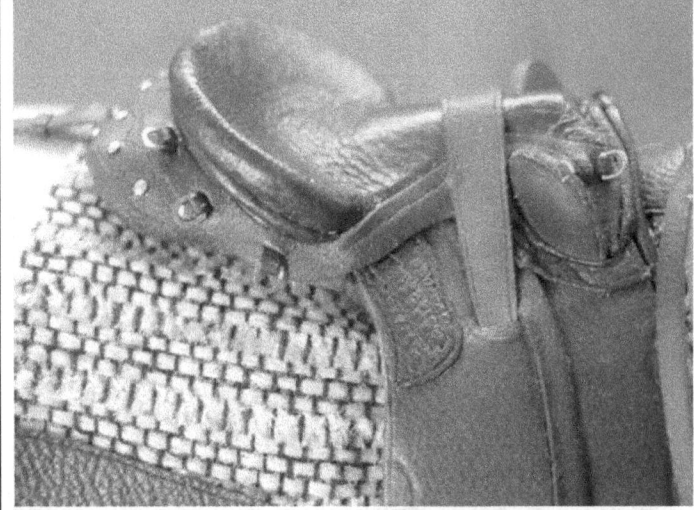

Work the seat area before the glue sets up to get a depression. Use six instead of four D-rings in the skirt flaps. Omit the rear poleys but emboss the stitching.

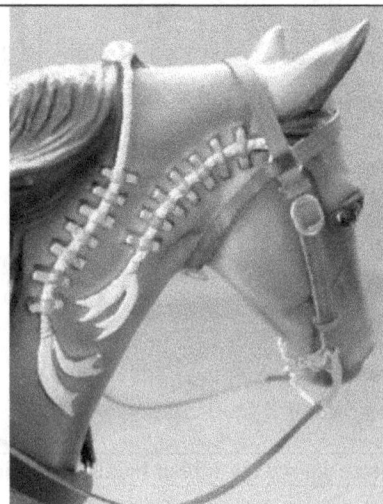

Western bit with Western reins on a Barco Bridle.

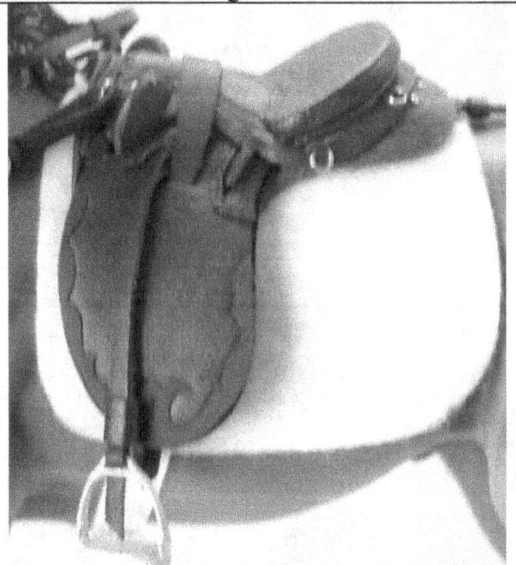

For a two-toned saddle: here I used black for the skiver and tan for the 2-oz leather.

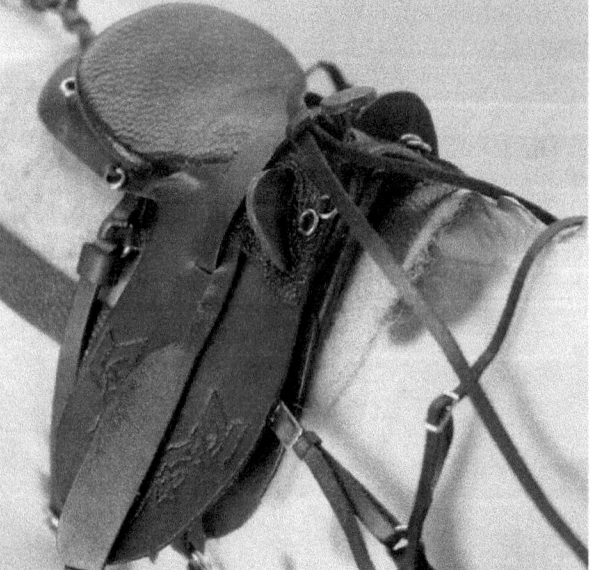

How about a Western version with front poley panels? Also, I added a neck strap to the Western breast collar.

Putting It All Together

NAN qualified entry
Lil' Girl's Dream

    Here's one of my entries for endurance riding. The class was Other Performance in the Other Performance division. I was going for an overall champion with this model so every ribbon mattered. The binoculars I found in the shadow box section of my local craft store. The canteen was an old toy from my childhood that I covered in aluminum foil and then felt. Pattern and assembly schematics for the saddlebags are located in the back of the book. For the class, she also had leg protection made from a thin Velcro wrap - hooks on one side and loops on the other. She received 2nd place for the class and Reserve Grand Performance Champion for the show.

Saddle Pieces and Patterns

Make photocopies, or scan into your computer and print as needed. If you prefer the trace method, you cut out the patterns; laminate or use clear tape to protect the pattern pieces for multiple use. Since you might have a problem getting copies, due to copyright laws, show them this page where I give permission for you to make copies of the pattern piece pages (pattern piece pages only!) at the end of this book. Please do NOT share copies of the patterns with others!

There are five different sizes of plastic models between Breyer and Peter Stone horses (and there may be more). Just in case you want to reduce the traditional size pattern, I have provided the reduction ratios. You can use your graphics software or a copy machine with reduction capabilities. I have given each size a name and letter/number indicator for reference. Mark your copies accordingly to help you remember. Always test for size using the skirt or seat length before making a saddle for a particular mold. I found this out the hard way, completing a saddle only to find it way too large for the model I designed it for!

Reduction Ratios

- T1 (Traditional 1 or 1:9 scale) – Traditional size Breyer and Peter Stone horses. For this pattern, also includes most Arabians. T1 is 100%.
- T2 (Traditional 2) – Traditional size ponies and some short backed Traditional models. T2 is 85% of T1 – reduce by 15%.
- C1 (Classic 1) – Models Breyer classifies as Traditional that are in fact, a size between Traditional and Classic models such as Scamper. Also, for some of the Traditional size ponies. C1 is 75% of T1 – reduce by 25%.
- C2 (Classic 2 or 1:12 scale) – The standard classic size models such as Kelso and Ruffian. C2 is 70% of T1 – reduce by 30%.
- P1 (Pebbles1) – Breyer Little Bits/Paddock Pals and Stone Horses Pebbles models. Although the chests on the Stone models are deeper, the backs are basically the same size. P1 is 50% of T1 – reduce by 50%.
- Stablemates (S1) are too small for me. It would have to be made entirely of 1oz-leather. (You might want to invest in a microscope as well!) You are welcome to give it a try. S1 (Stablemates) is 25% of T1 – reduce by 75%.

Blocking Form Tool

Cut 1 each - scrap 7/8 oz or 1/8" thick leather

Base

Top

Cut out center on straight lines. Use dotted lines to create guide lines.

Tree form

Seat top

Seat back

Pieces Common to Both Versions

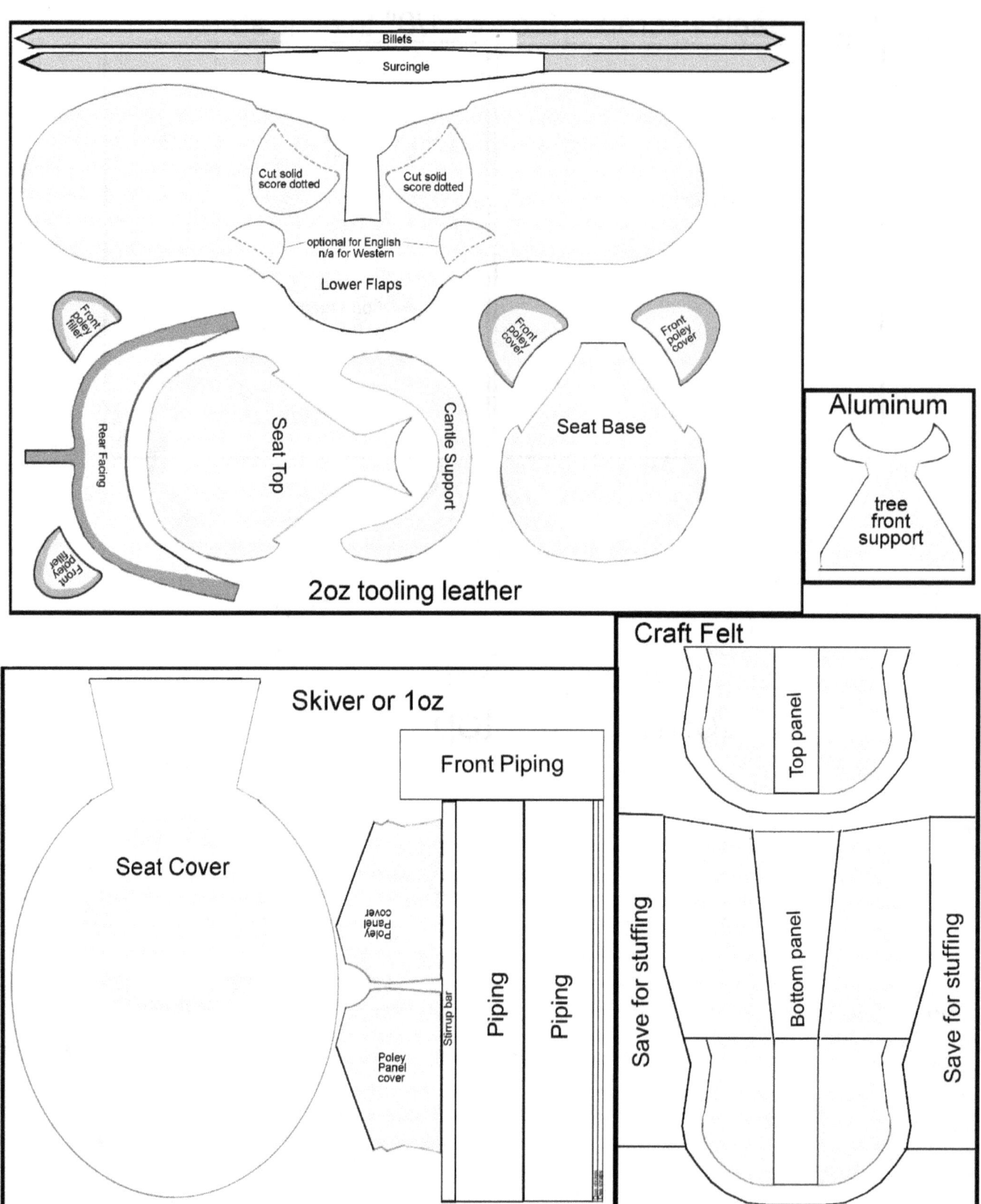

English Version Pieces

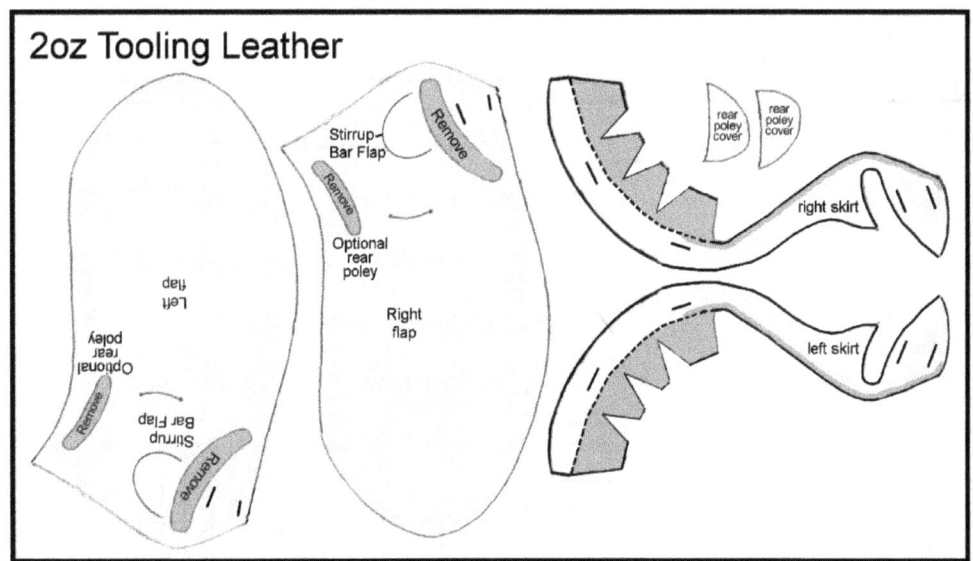
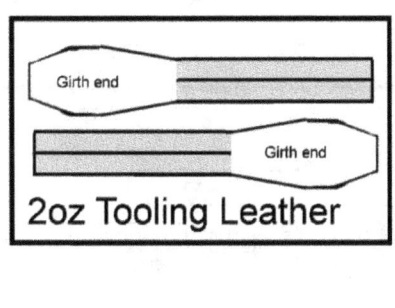
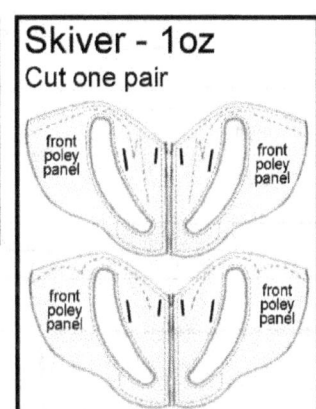
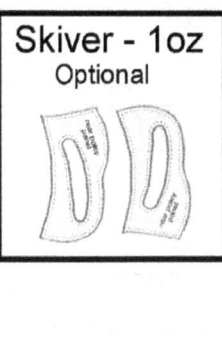
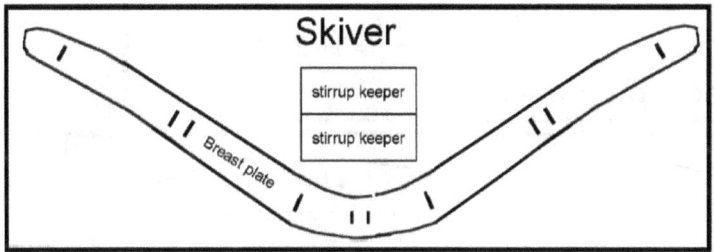

Western Version Pieces

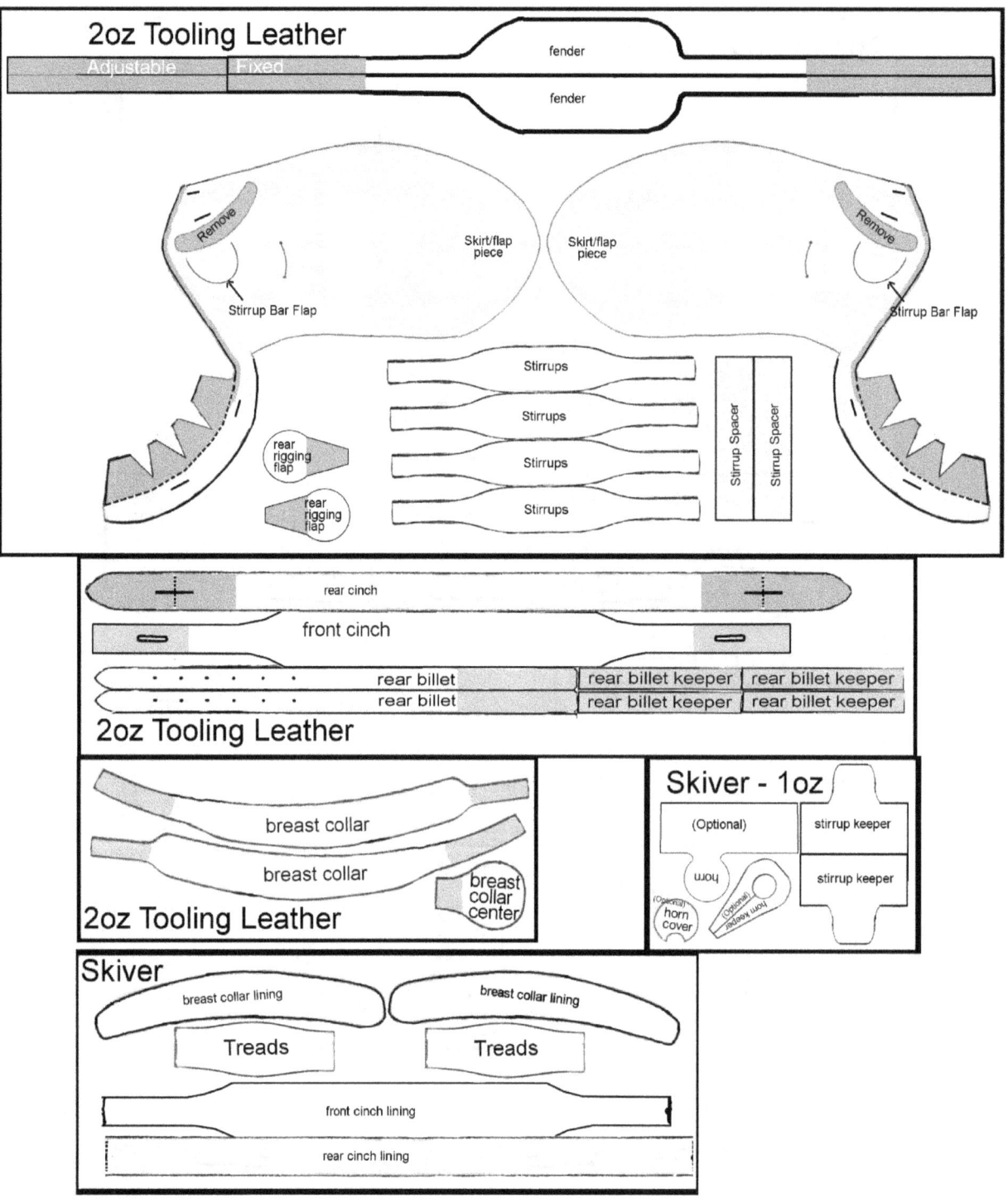

Saddle Blankets

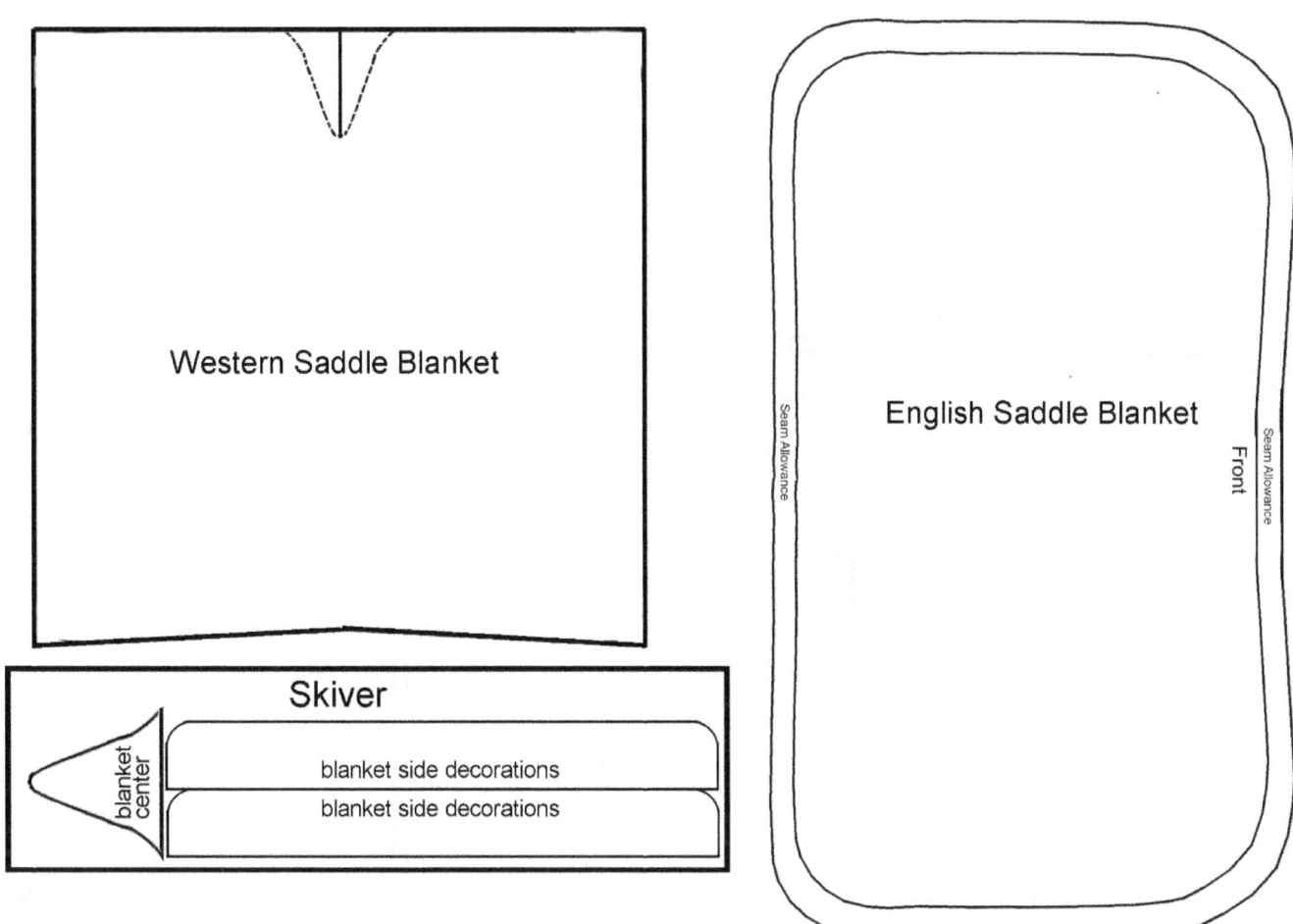

Faux Stitching and Tooling – English Skirts and Flaps, Rear Facing

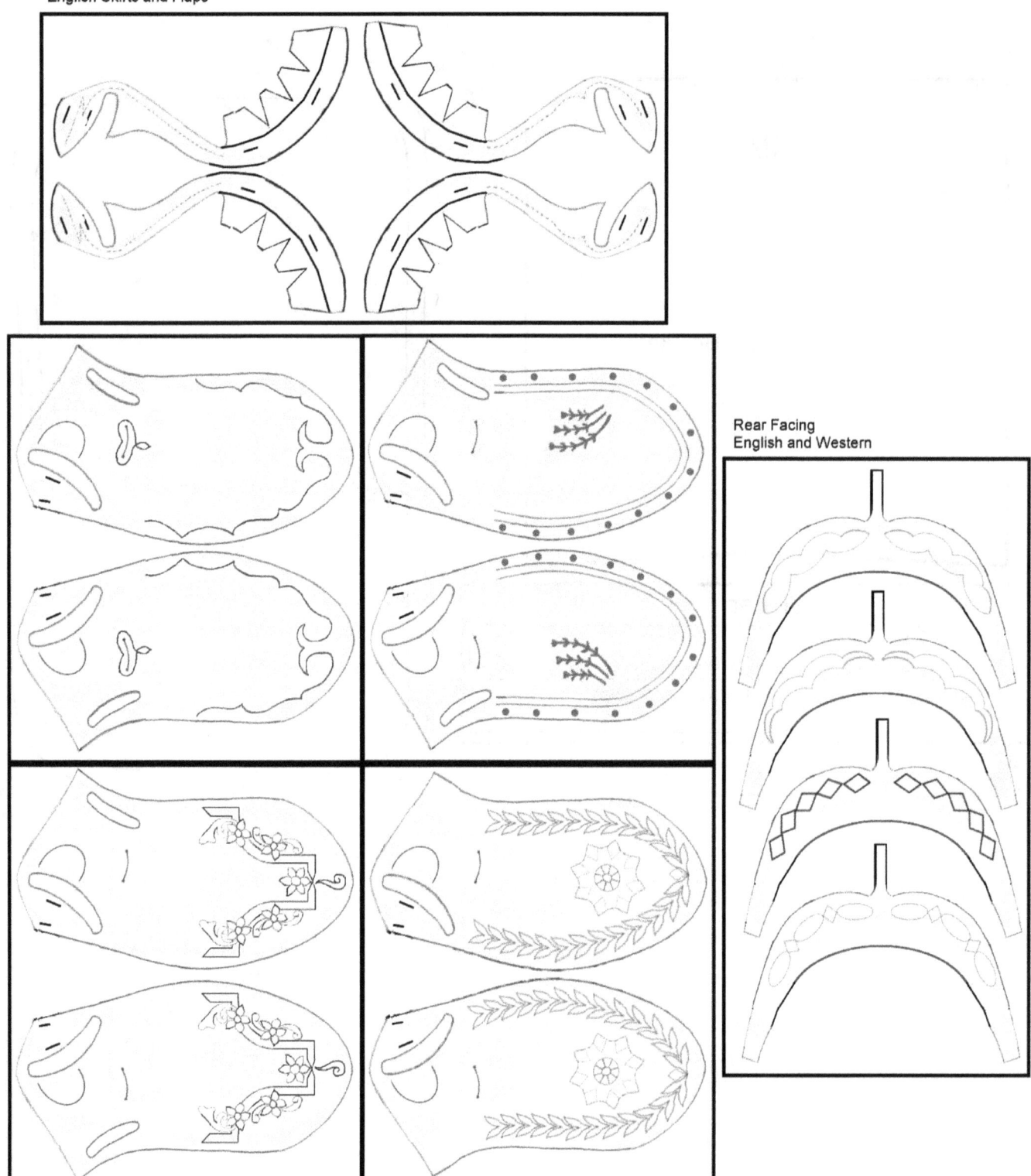

Faux Stitching and Tooling – Western Skirt/Flaps

Western Skirt/Flaps

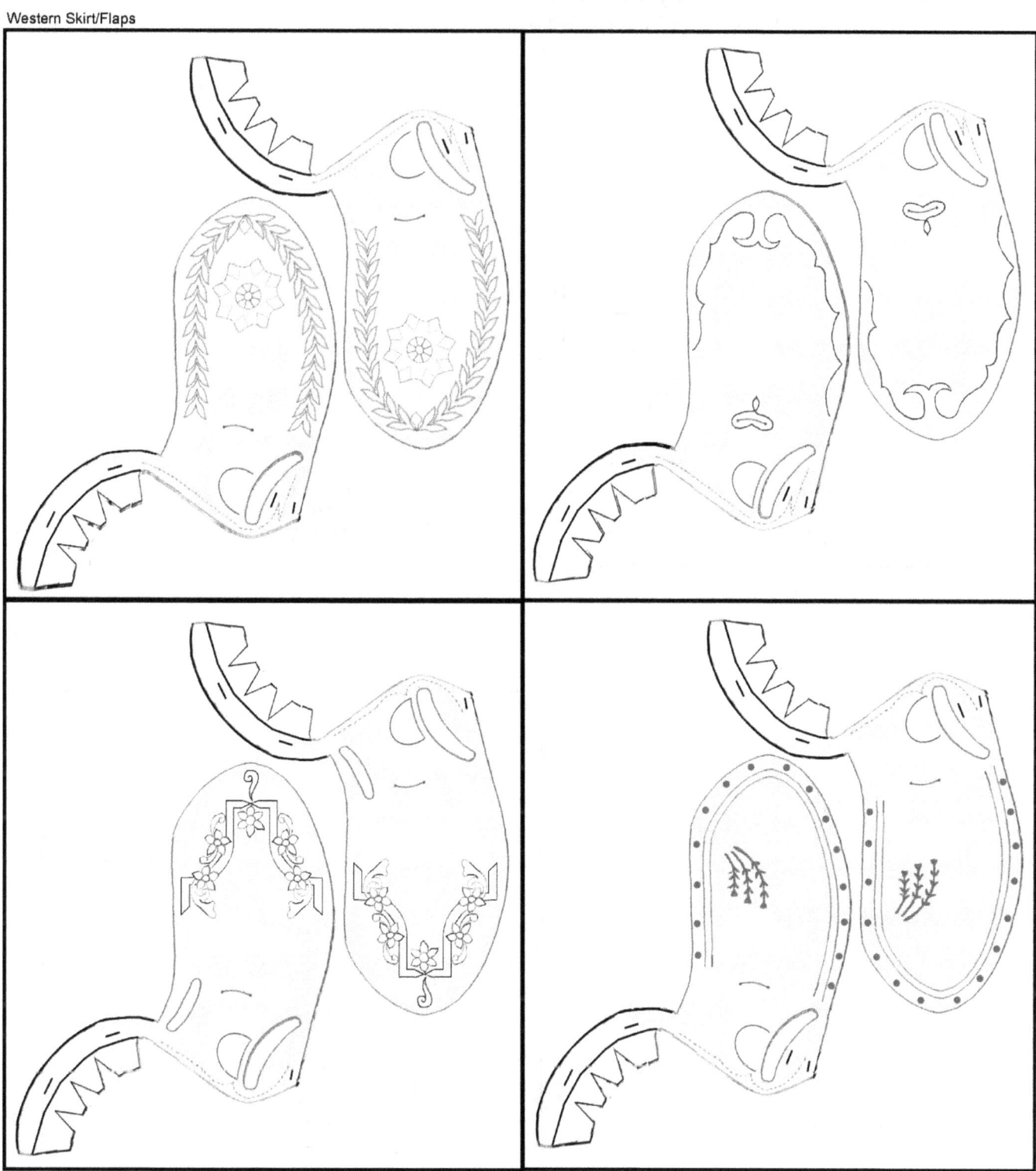

Saddle Bag Pattern and Assembly Schematic

I tried to make a step-by-step with pictures but it was impossible to get inside the bag. Rather than avoid sharing the pattern, I created this schematic. It's not my forte. But it was the best way I could think of to show the assembly process. I hope this works.

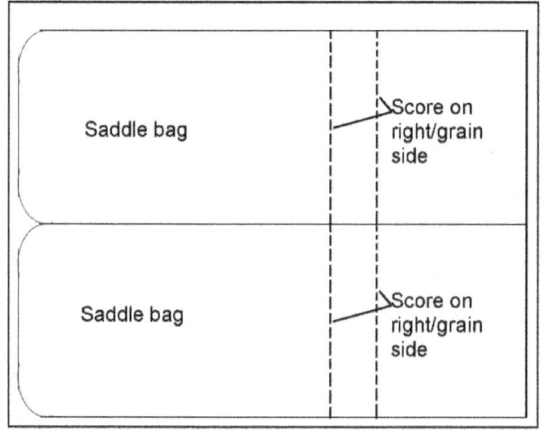

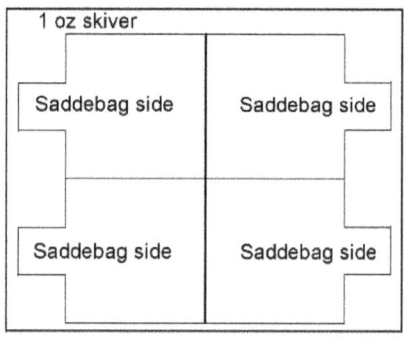

Leather lace 1/16" or 1/8" width

4 buckles to match leather lace width
two 4" lengths for front buckle straps
two 2" lengths for rear buckle straps
two 3' lengths for rear buckle connector straps

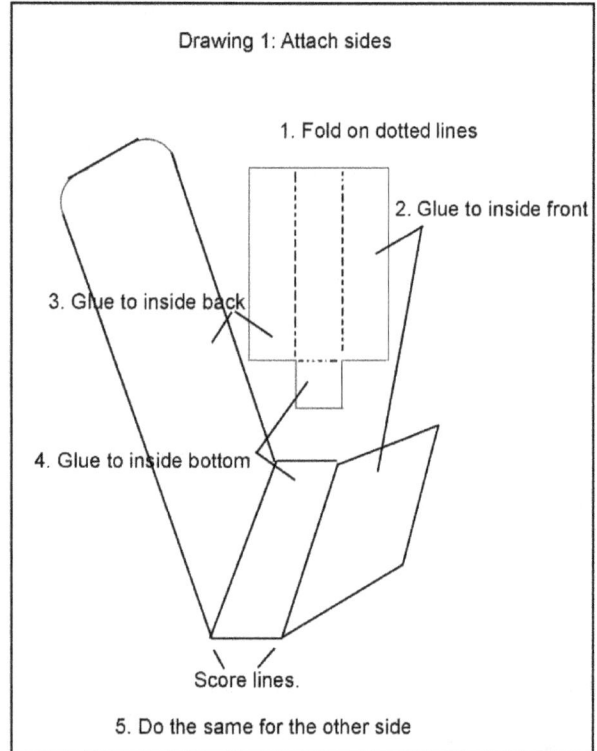

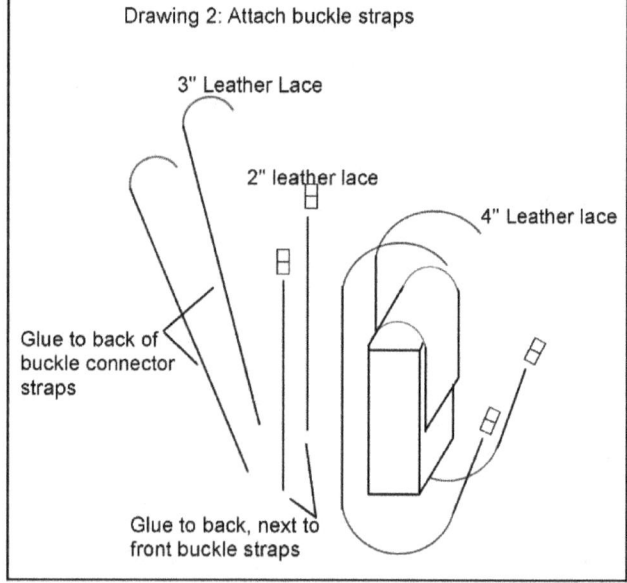

www.ingramcontent.com/pod-product-compliance
Lightning Source LLC
Chambersburg PA
CBHW080226170426
43192CB00015B/2763